MW01012045

The Nalco Guide
to Cooling Water System
Failure Analysis

The Nalco Guide to Cooling Water System Failure Analysis

Nalco Chemical Company

Authored by

Harvey M. Herro

Robert D. Port

McGraw-Hill, Inc.
New York St. Louis San Francisco Auckland Bogota
Caracas Lisbon London Madrid Mexico City Milan
Montreal New Delhi San Juan Singapore
Sydney Tokyo Toronto

Library of Congress Cataloging-in-Publication Data

Herro, Harvey M.
    The Nalco guide to cooling water system failure analysis / Nalco
Chemical Company ; authored by Harvey M. Herro, Robert D. Port.
        p.   cm.
    Includes bibliographical references and index.
    ISBN 0-07-028400-8
    1. Cooling towers.   2. Heat exchangers—Corrosion.   I. Port,
Robert D.   II. Nalco Chemical Company.   III. Title.
TJ563.H47   1993
621.1'97—dc20   92-22428
                                                                    CIP

Copyright © 1993 by McGraw-Hill, Inc. All rights reserved. Printed in
the United States of America. Except as permitted under the United
States Copyright Act of 1976, no part of this publication may be reproduced or distributed in any form or by any means, or stored in a data
base or retrieval system, without the prior written permission of the
publisher.

891011 KGPKGP 054321

ISBN   0-07-028400-8

*The sponsoring editor for this book was Robert W. Hauserman, the
editing supervisor was David E. Fogarty, and the production supervisor
was Suzanne W. Babeuf. It was set in Century Schoolbook by North
Market Street Graphics.*

Printed and bound by Kingsport Press.

Information contained in this work has been obtained by McGraw-
Hill, Inc. from sources believed to be reliable. However, neither
McGraw-Hill nor its authors guarantee the accuracy or complete-
ness of any information published herein, and neither McGraw-
Hill nor its authors shall be responsible for any errors, omissions, or
damages arising out of use of this information. This work is published with the understanding that McGraw-Hill and its authors
are supplying information but are not attempting to render engineering or other professional services. If such services are required,
the assistance of an appropriate professional should be sought.

To our families for their patience, understanding, and support.

Contents

Preface

Cooling water system corrosion causes immediate and delayed problems. Difficulties spread from a failure like ripples from a pebble thrown into a pool. A single failure may force an unscheduled outage, redirect worker efforts, contaminate product, compromise safety, increase equipment expense, violate pollution regulations, and decrease productivity.

The closer one is to the failure, the more its direct effects are apparent. The cumulative effects of failure are often overlooked in the rush to "fix" the immediate problem. Too often, the cause of failure is ignored or forgotten because of time constraints or indifference. The failure or corrosion is considered "just a cost of doing business." Inevitably, such problems become chronic; associated costs, tribulations, and delays become ingrained. Problems persist until cost or concern overwhelm corporate inertia. A temporary solution is no longer acceptable; the correct solution is to identify and eliminate the failure. Preventative costs are almost always a small fraction of those associated with neglect.

The solution to any cooling water system failure begins with a thorough understanding of the cause. Careful, meticulous investigation will almost always reveal the failure source and any attendant accelerating factors. Ideally, potential problems will be identified before failure. The identification of cooling water system potential problems begins with a knowledge of how to recognize such problems. Knowing where particular forms of damage might occur, what damage looks like, how critical factors influence damage, and, most importantly, how such problems can be eliminated are the objectives of this book.

This book follows the format used in *The Nalco Guide to Boiler Failure Analysis,* also authored by Robert D. Port and Harvey M. Herro, copyright 1991 by McGraw-Hill, Inc. Each chapter is divided into eight sections, giving specific information on damage:

General Description

Locations

Critical Factors

Identification

Elimination

Cautions

Related Problems

Case Histories

All areas of the cooling water system where a specific form of damage is likely to be found are described. The corrosion or failure causes and mechanisms are also described. Especially important factors influencing the corrosion process are listed. Detailed descriptions of each failure mode are given, along with many common, and some not-so-common, case histories. Descriptions of closely related and similarly appearing damage mechanisms allow discrimination between failure modes and avoidance of common mistakes and misconceptions.

Many sources contain scattered information concerning cooling water system corrosion and defects, and many literature studies describe corrosion processes and mechanisms from a predominantly theoretical viewpoint. Until now, however, no source discusses cooling water system corrosion with emphasis on identification and elimination of specific problems. Much of the information in this book is unique; every significant form of attack is thoroughly detailed. Color photos illustrate each failure mechanism, and case histories further describe industrial problems.

Visual inspection techniques are stressed as the most important tools used to study failures. This text is not a substitute for rigorous failure analysis conducted by experts, but it will help the reader identify and eliminate many cooling water system problems. Still, on occasion, the experienced, skilled, failure analyst using sophisticated analytical techniques and specialized equipment may be required to solve complex or unusual problems. Common sense, appropriate experience, and systematic investigation are, however, often superior to the more elaborate, but less effective, techniques used by some.

Acknowledgments

Thanks are extended to Nalco Chemical Company for its support during this work. Our colleagues James J. Dillon and James G. Tomei are the recipients of special appreciation for their contributions of selected photos and case histories. Thanks are extended to Michael J. Danko for his contribution of selected photos and aid in sample preparation. We are indebted to Richard W. Cloud for his excellent scanning electron microscopy work. Warmest thanks to Ted R. Newman, our mentor. Thanks are extended to Dr. William K. Baer for his support and guidance during this work. Credit is due James E. Shannon and Bruce Thompson for their manuscript review. Special thanks to Christopher Wiatr for his review of Chap. 6, "Biologically Influenced Corrosion." Constance Szewczyk deserves special thanks for aid in manuscript preparation. Thanks also to Alice LaMaide for her secretarial assistance. Finally, sincere gratitude is given to Pamela J. Entrikin for final manuscript edits, cover design direction, and assistance in coordinating the production of this book with McGraw-Hill.

Harvey M. Herro
Robert D. Port

ABOUT THE AUTHORS

HARVEY M. HERRO obtained a B.S. in Physics from Marquette University and a Ph.D. in metallurgical engineering from Iowa State University. Dr. Herro, who has been with Nalco since 1982, holds patents for corrosion monitoring using chemical techniques, is the author of numerous papers on corrosion and failure analysis, and is a frequent lecturer. He is an active member in the National Association of Corrosion Engineers (NACE) and the American Society for Metals (ASM). Dr. Herro currently chairs Task Group T-3A-12 Underdeposit Corrosion at NACE and is an active member of many of its committees. At Nalco, he leads the Nalco Corrosion Detection Group and conducts training on recognizing forms of industrial corrosion.

ROBERT D. PORT is a metallurgical engineer with more than twenty years' experience in failure analysis, at both Nalco Chemical Company and at independent metallurgical laboratories. Mr. Port is the author of numerous papers on failure analysis, and frequently speaks to technical societies and industrial groups. He is a member of NACE and is active on its committee for Failure Analysis in Steam Generating Systems and its committee on Corrosion in Steam Generating Systems. Mr. Port attended the University of Illinois where he received his B.S. in Metallurgical Engineering.

Mr. Port and Dr. Herro have made presentations at professional societies such as the Electric Power Research Institute (EPRI), NACE, ASM, the American Boiler Manufacturer's Association (ABMA), and the American Chemical Society (ACS). Together, the authors have completed more than 4700 failure analyses.

Nalco Chemical Company, headquartered in Naperville, Illinois, has worldwide sales of over one billion dollars.

Cooling Water System Design and Operation

Cooling systems suffer many forms of corrosion and failure. The diversity of attack is caused by differences in cooling water system design, temperature, flow, water chemistry, alloy composition, and operation. An almost endless variation of process stream chemistries may be involved in cooling water systems. Refinery and chemical process industries can employ hundreds of heat exchangers at a single plant, each with a different process stream chemistry. Hence, portions of the system contacting water (and sometimes steam) will mostly be discussed in this text. As the reader will note, the variety of problems encountered on waterside surfaces are formidable enough.

Types of Systems

Cooling water systems are either open or closed, and water flow is either once-through or recirculating. The three basic types of cooling water systems are once-through, closed recirculating (nonevaporative), and open recirculating (evaporative). Each is shown schematically in Figs. 1.1 through 1.3.

True closed systems neither lose nor gain water during service. Open systems, however, must have water added to make up for losses.

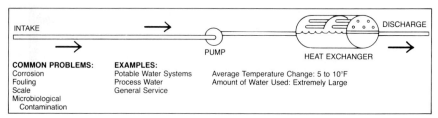

Figure 1.1   Schematic of once-through cooling system.

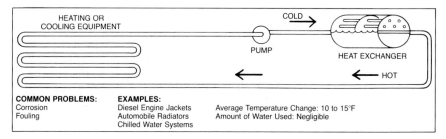

Figure 1.2   Schematic of closed recirculating cooling system.

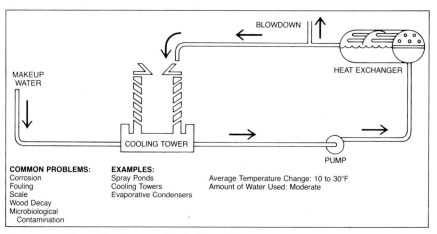

Figure 1.3   Schematic of open recirculating cooling system.

Open recirculating systems employing cooling towers and spray ponds allow the dissipation of enormous heat loads while limiting the amount of water consumed. An open recirculating water loop joined to a closed loop used in an air-conditioning system is diagrammed in Fig. 1.4. Typical hyperbolic cooling towers in a utility are shown in Fig. 1.5. Once-through cooling takes water from a plant supply, passes it

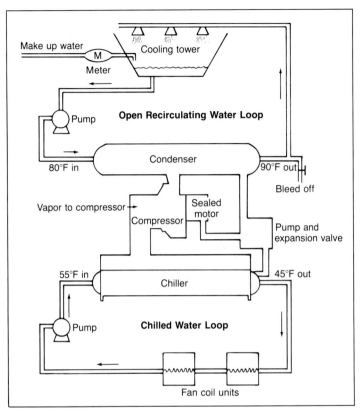

Figure 1.4 Schematic of all components of a complete air-conditioning system. [*Fig. 37.8, The Nalco Water Handbook, 1st ed. (1979), reprinted with permission from McGraw-Hill, Inc.*]

through a cooling system, and finally sends it to a receiving body of water. Closed recirculating system operation is obvious, since no water is added or lost.

Closed and open systems experience forms of attack related to the amount of water added to the system. In closed systems (nonevaporative) where water loss is low, the total waterborne material entering the system is limited. Thus, deposited minerals accumulate at a much slower rate than in systems in which large amounts of makeup water are added. Open recirculating (evaporative) and once-through systems are exposed to large quantities of solutes, suspended solids, and biological materials. As a consequence, fouling and associated corrosion are generally more significant in open systems than in true closed systems. The amount of deposit is not always proportional to the damage

Figure 1.5   Hyperbolic towers cooling condenser water in a utility station. [*Fig. 38.10, The Nalco Water Handbook, 1st ed. (1979), reprinted with permission from McGraw-Hill, Inc., Courtesy of The Marley Company.*]

produced, however. Often, closed systems are used in critical cooling operations where even a small amount of deposit cannot be tolerated (Fig. 1.6).

Cooling System Equipment

Cooling water system designs and equipment vary widely depending upon application. Included are heat exchangers, transfer piping, pumps, cooling tower components, and valves. By far, the greatest variety of designs involves heat exchangers.

Heat exchangers have two common designs: shell-and-tube and plate-and-frame designs. Shell-and-tube heat exchangers are very common. A typical shell-and-tube design is shown in Fig. 1.7. The most frequently affected component in the shell-and-tube exchanger is the tubes. Corrosion fatigue, stress-corrosion cracking, erosion-corrosion, underdeposit attack, dealloying, oxygen corrosion, and many other forms of wastage frequently occur. Tube sheets, baffles, and water boxes also may be damaged by any of the aforementioned mechanisms.

Plate-and-frame exchangers transfer heat by passing cooling fluids and process fluids between large corrugated panels. Crevices exist between closely spaced panels that stimulate localized attack. Plates

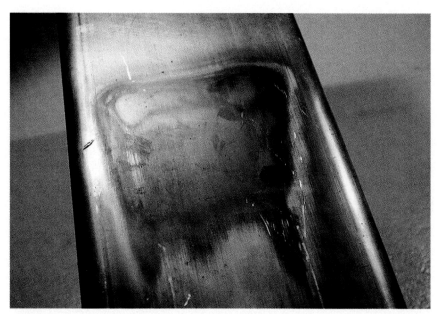

Figure 1.6 Dark oxide and deposit lobes on a copper continuous caster mold from a steel-making operation. Since heat transfer is high, even small amounts of deposit are unacceptable.

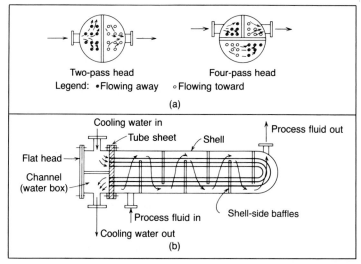

Figure 1.7 Simple detail of shell-and-tube heat exchanger. The water box may be designed for as many as eight passes, and a variety of configurations of shell-side baffles may be used to improve heat transfer. (a) Several water box arrangements for tube-side cooling. (b) Assembly of simple two-pass exchanger with U-tubes. [*Fig. 38.2, The Nalco Water Handbook, 1st ed. (1979), reprinted with permission from McGraw-Hill, Inc.*]

are often exposed to a highly oxygenated environment and, consequently, are fabricated of stainless steel. Stress-corrosion cracking—
caused by the combined actions of high residual forming stresses, high
operating temperatures, and evaporative concentration of chlorides—
frequently occurs.

Alloy Choice

Virtually all commercially available alloys have been used in cooling
water systems. A partial listing of the more common alloys and uses is
given in Table 1.1. Often, materials used for heat exchanger and other
cooling applications are either innately noble or show a strong tendency to passivate in a particular cooling water environment. (Corrosion spontaneously decreases when exposed to a given environment.)
Of course, process streams must also be tolerated by the chosen alloy.

TABLE 1.1   Common Alloys Used in Cooling Water Systems

Alloy	Common uses
Carbon steel, low-alloy steels	Transfer lines, heat exchanger shells, baffles, pump components, heat exchanger tubing, fan blades and shrouds, valves, screens, fasteners
Cast iron	Pump housings and impellers, valves, plumbing fixtures, large diameter pipe
Galvanized steel	Cooling tower components, fan blades and shrouds, transfer pipes, plumbing fixtures
Bronzes	Bearings, pump impellers, special-purpose tubing, heat exchanger tubing, screens
Copper and brasses	Heat exchanger tubing, bearings, valve components, gaskets, brewing equipment
Cupronickel	Heat exchanger tubing
Stainless steels 300 series (304, 304L, 310, 316, 316L, etc.)	Plate-and-frame exchangers, heat exchanger tubing, pump impellers and housings, heat exchanger shells, brewing equipment, drying equipment, oil refinery still tubes, pulp and paper equipment
Duplex	Stress-corrosion-cracking-resistant tubing
400 series	Special-purpose heat exchanger tubing, plate-and-frame heat exchangers
Aluminum	Heat exchanger tubing, transfer piping
Molybdenum	Furnace electrodes used to melt siliceous compounds
Nickel	Special-purpose heat exchangers, caustic handling
Titanium	Fresh and sea water condenser tubing, severe corrosion condition equipment

Additionally, heat transfer is important, and economics is always a vital consideration.

Many shell-and-tube condensers use copper alloy tubes, such as admiralty brasses (those containing small concentrations of arsenic, phosphorus, or antimony are called inhibited grades), aluminum brasses, and cupronickel; austenitic stainless steel and titanium are also often used. Utility surface condensers have used and continue to use these alloys routinely. Titanium is gaining wider acceptance for use in sea water and severe service environments but often is rejected based on perceived economic disadvantages.

The single most important factor in choosing a replacement alloy for condenser tubing is service experience in a specific condenser. Although laboratory experience can be valuable in alloy selection, it should never be the sole reason for alloy choice. Past corrosion should be identified and condenser service conditions reviewed. If possible, a few new alloy tubes should be used to replace existing tubes in the condenser for a test period. After the test, replacement tubes should be carefully examined for signs of attack. Nature can be perverse. Although old corrosion problems may be solved by an alloy substitution, new troubles may occur that are as bad or worse than original problems. There is no totally acceptable substitute for experience.

Concentration Cell Corrosion

Introduction

The three major forms of concentration cell corrosion are crevice corrosion, tuberculation, and underdeposit attack. Each form of corrosion is common in cooling systems. Many corrosion-related problems in the cooling water environment are caused by these three forms of wastage. The next three chapters—Chap. 2, "Crevice Corrosion," Chap. 3, "Tuberculation," and Chap. 4, "Underdeposit Corrosion"— will discuss cooling water system corrosion problems.

Attack associated with nonuniformity of the aqueous environments at a surface is called concentration cell corrosion. Corrosion occurs when the environment near the metal surface differs from region to region. These differences create anodes and cathodes (regions differing in electrochemical potential). Local-action corrosion cells are established, and anodic areas lose metal by corrosion. Shielded areas are particularly susceptible to attack, as they often act as anodes (Fig. 2.1). Differences in concentration of dissolved ions such as hydrogen, oxygen, chloride, sulfate, etc. eventually develop between shielded and nearby regions.

Shielded areas may be produced by design or accident. Crevices caused by design occur at gaskets, flanges, washers, bolt holes, rolled tube ends, contact points in plate-and-frame heat exchangers, threaded joints, riveted seams, overlapping screen wires, lap joints, beneath insulation, and anywhere close-fitting surfaces are present. Occluded regions are formed beneath deposits and below accumulations of biological materials. Coatings may partially exfoliate, producing crevices.

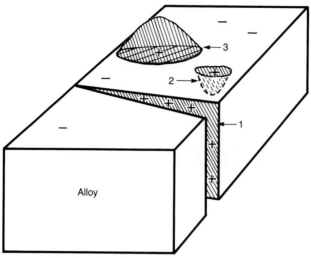

Figure 2.1   Region of local action cells in an alloy immersed in an electrolyte. 1. Crevice, 2. pit, 3. deposit. Shaded regions are anodic (+) to surrounding metal (−).

Virtually all metallurgies are attacked if the environment is sufficiently aggressive. Alloys that strongly passivate in oxygenated waters (those readily forming protective oxide layers) such as stainless steels are more susceptible to localized concentration cell corrosion than alloys with weaker passivation tendencies. Chloride, sulfate, thiosulfate, atomic hydrogen, or other potentially corrosive ions weaken protective oxide layers, accelerating attack. Regions not shielded remain protected, and corrosion is focused into localized areas, sometimes producing deep pits. Similarly, aluminum and even titanium are attacked under suitably severe conditions. Weakly passivating metals are also affected. However, attack is likely to be more general and pitting less pronounced. Coupling dissimilar metallurgies may magnify wastage.

In each form of attack, solute concentration differences arise primarily by diffusion-related processes. As a consequence, stagnant conditions may promote attack, since concentration gradients near affected areas are reduced by flow and these concentration gradients supply the energy that drives diffusion. Similarly, high concentrations of dissolved species increase attack. Elevated temperature usually stimulates attack by increasing both diffusion and reaction rates.

Exceptions to these rules of thumb abound, however. For example, although diffusion is faster at elevated temperatures, dissolved oxygen concentration may be lower. Convective transport is stimulated by high temperatures, but increased turbulence and thermal stresses may dislodge deposits or promote convective flow into and out of crevices.

Finally, pitting may be viewed as a special form of concentration cell corrosion. Most alloys that are susceptible to crevice corrosion also pit. However, many metals may pit but not show crevice attack. Further, although sharing many common features with concentration cell corrosion, pitting is sufficiently different to warrant a separate categorization.

Crevice Corrosion

General Description

The crevice must be filled with water. Surfaces adjacent to the crevice must also contact water. Without liquid filling the crevice, corrosion cannot occur. The crevice typically is a few thousandths of an inch wide. Corrosion may extend up to several feet into a crevice mouth in high conductivity waters.

The crevice geometry and normally occurring corrosion combine to produce accelerated attack in the shielded region, a so-called autocatalytic process. Initially, corrosion in oxygenated water of near neutral pH occurs by Reactions 2.1 and 2.2:

Anode: $$M \rightarrow M^{+n} + ne^-$$ (2.1)

Cathode: $$O_2 + 2H_2O + 4e^- \rightarrow 4OH^-$$ (2.2)

These reactions are shown schematically near a crevice in Fig. 2.2. Many other reactions may occur at anodes and cathodes, but Reactions 2.1 and 2.2 predominate on carbon steel, for example, in near neutral pH, oxygenated water.

Eventually oxygen becomes depleted in the crevice. Replenishment of oxygen by convective transport cannot occur since the crevice is too tight to allow water to move in and out freely. Also, oxygen diffusion

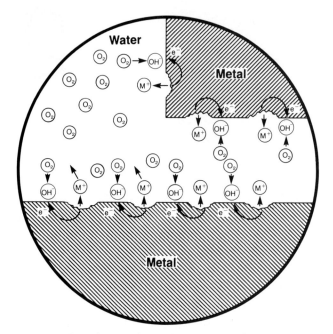

Figure 2.2   Initial stage of corrosion near a crevice.

into the crevice is too slow to replace oxygen as fast as it is consumed during corrosion. Actually, diffusion of oxygen in water near room temperature is relatively rapid; ions move about one centimeter per hour. The cross-sectional area of the crevice mouth relative to the internal crevice area is small, however. A typical crevice might have an internal area five or six orders of magnitude greater than the crevice mouth area. Hence, the total amount of oxygen transported is insufficient to replace oxygen consumed internally.

Oxygen concentration is held almost constant by water flow outside the crevice. Thus, a differential oxygen concentration cell is created. The oxygenated water allows Reaction 2.2 to continue outside the crevice. Regions outside the crevice become cathodic, and metal dissolution ceases there. Within the crevice, Reaction 2.1 continues (Fig. 2.3). Metal ions migrating out of the crevice react with the dissolved oxygen and water to form metal hydroxides (in the case of steel, rust is formed) as in Reactions 2.3 and 2.4:

$$4M^+ + 4e^- + O_2 + 2H_2O \rightarrow 4MOH \downarrow \qquad (2.3)$$

In the case of steel,

$$2Fe^{++} + 4e^- + O_2 + 2H_2O \rightarrow 2Fe(OH)_2 \downarrow \qquad (2.4)$$

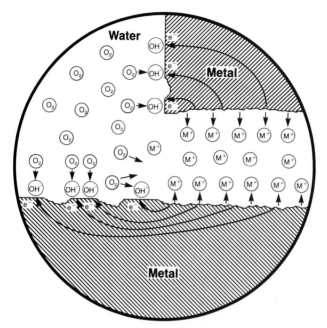

Figure 2.3 Later stage as in Fig. 2.2 when an oxygen concentration cell is established between regions inside and outside the crevice. Oxygen has been depleted inside the crevice.

As metal ion concentration increases in the crevice, a net positive charge accumulates in the crevice electrolyte. This attracts negatively charged ions dissolved in the water. Chloride, sulfate, and other anions spontaneously concentrate in the crevice (Figs. 2.4 and 2.5). Hydrolysis produces acids in the crevice, accelerating attack (Reactions 2.5 and 2.6). Studies have shown that the crevice pH can decrease to 2 or less in salt solutions having a neutral pH.

$$M^+Cl^- + H_2O \rightarrow MOH \downarrow + H^+Cl^- \tag{2.5}$$

$$M_2^+SO_4^= + 2H_2O \rightarrow 2MOH \downarrow + H_2^+SO_4^= \tag{2.6}$$

In sea water with a pH of 8, crevice pH may fall below 1 and chloride concentration can be many times greater than in the water. The crevice environment becomes more and more corrosive with time as acidic anions concentrate within. Areas immediately adjacent to the crevice receive ever-increasing numbers of electrons from the crevice. Hydroxyl ion formation increases just outside the crevice—locally increasing pH and decreasing attack there (Reaction 2.2). Corrosion inside the crevice becomes more severe with time due to the spontaneous concentration of acidic anion. Accelerating corrosion is referred

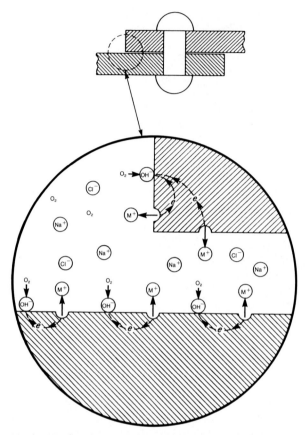

Figure 2.4 Crevice corrosion—initial stage in oxygenated water containing sodium chloride. (*Courtesy of Mars G. Fontana and Norbert D. Greene, Corrosion Engineering, McGraw-Hill Book Company, New York City, 1967.*)

to as *autocatalytic* due to the spontaneous corrosion increase that feeds upon itself.

Waterline attack

A special form of crevice attack can occur at a waterline or at the edges of water droplets. At the water surface, a meniscus region is present where surface tension causes water to climb up the metal surface it contacts. In effect, a crevice is formed between the air-liquid and liquid-metal interface at the meniscus. Oxygen concentration is high at the meniscus due to the greater accessibility of this region to the air. The meniscus region becomes cathodic to the adjacent metal surface. Corrosion occurs just below the meniscus, and chloride, if present, is

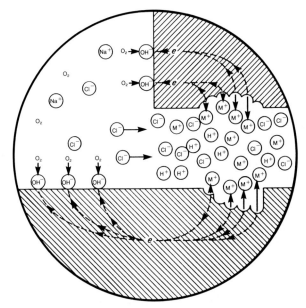

Figure 2.5 Crevice corrosion—final stage in oxygenated water containing sodium chloride. (*Courtesy of Mars G. Fontana and Norbert D. Greene, Corrosion Engineering, McGraw-Hill Book Company, New York City, 1967.*)

attracted to the anodic regions. Rust (on steel) or other metal hydroxides usually precipitate nearby. This process is shown in Fig. 2.6.

A (316) stainless steel beaker that failed by waterline attack is shown in Fig. 2.7. The beaker contained water into which a chlorinated biocide tablet had been placed. The 0.020-in.-thick (0.051 cm) beaker material perforated near the waterline in less than 40 hours. This corresponds to a minimum localized attack rate of 4.4 in. (11 cm) per year.

Locations

As the name implies, crevice corrosion occurs between two surfaces in close proximity, such as a crack. Table 2.1 gives a partial listing of common crevice corrosion sites in cooling water systems.

Any region where two surfaces are loosely joined or come into close proximity qualifies as a crevice site as long as water may enter. Partially exfoliated coatings may produce crevices. Surfaces may be metallic or nonmetallic. Usually, however, at least one surface is metallic. Crevices commonly are present at gaskets, flanges, washers, bolt holes, rolled tube ends, incompletely fused welds, contact points in plate-and-

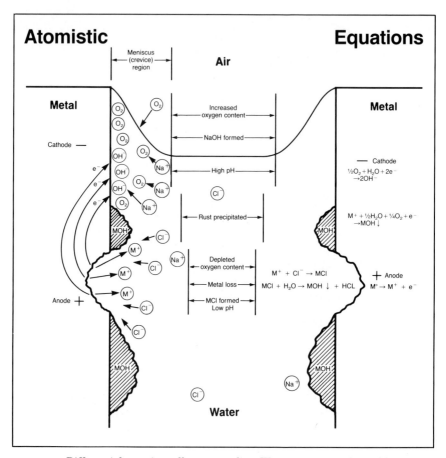

Figure 2.6   Differential aeration cell at a waterline. Water contains sodium chloride.

frame heat exchangers, threaded couplings, riveted seams, screen wires, lap joints, beneath insulation, and anywhere surfaces abut.

Critical Factors

Passivating metals

Metals that produce protective oxide layers (such as stainless steels) are especially susceptible to crevice attack. The reduced oxygen concentration in the crevice inhibits repair of the protective oxide film. This is especially true if acidic anions are present, which further retards oxide repair. Stainless steels containing molybdenum are usually less susceptible to attack.

Figure 2.7 Severe waterline attack in stainless steel beaker. The beaker contained a chlorinated biocide tablet in water over a weekend. Perforations occurred in 40 hours or less, giving a minimum corrosion rate of 4380 mils (11.1 cm) per year at the perforations.

TABLE 2.1   **Common Crevice Attack Sites in Cooling Water Systems**

Component	Location
Heat exchangers	
Shell and tube	Rolled ends at tube sheet
	Open welds at tube sheet
	Beneath deposits
	Water box gaskets
	Bolt holes, nuts, washers
	Baffle openings
	Disbonded water box linings
Plate and frame	Beneath gaskets
	Plate contact points
	Beneath deposits
Cooling tower	Threaded pipe joints
	Bolts, nuts, washers
	Partially exfoliated coatings
	Lap joints in sheet metal
	Between bushings and shafts on pumps
	Pump gaskets
Engine cooling	Gasketed surfaces
	Bolts, nuts, washers

Incubation period

There is often a period before corrosion starts in a crevice in passivating metals. This so-called incubation period corresponds to the time necessary to establish a crevice environment aggressive enough to dissolve the passive oxide layer. The incubation period is well known in stainless steels exposed to waters containing chloride. After a time period in which crevice corrosion is negligible, attack begins, and the rate of metal loss increases (Fig. 2.8).

The incubation period varies widely depending on such factors as crack morphology, water chemistry, and temperature. However, experience in a wide variety of cooling water environments has shown that many stainless alloys develop noticeable attack within 6 months of first being exposed to water. It is rare to see attack initiating many years after equipment commissioning unless service conditions change in the interim.

Acidic anion concentration

The amount of chloride, sulfate, thiosulfate, or other aggressive anions dissolved in water necessary to produce noticeable attack depends on many interrelated factors. Extraordinarily, if the water is quite aggressive, general corrosion may occur so rapidly outside the crevice that concentration differences cannot easily develop between the crevice interior and exterior. However, it is usually safe to assume that as the concentration of aggressive anions increases in solution, crevice attack is stimulated. Seawater chloride concentrations produce severe attack in most stainless crevices in a few weeks.

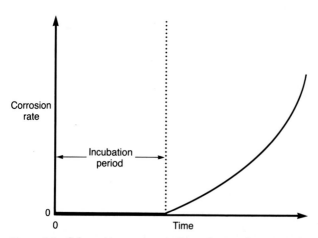

Figure 2.8 Schematic representation of corrosion rate as a function of time in a crevice in stainless steel exposed to chloride-containing water. The time before corrosion initiation is called the incubation period.

There is no limit at which the concentration of these ions in solution becomes "safe." However, as concentration decreases, severity of attack almost always decreases. Concentration may occur by evaporation in alternate wet-dry conditions.

Crevice morphology

The crevice shape markedly affects corrosion. Crevices so tight that water may not enter are entirely immune to attack. In misting environments or alternately wet-dry environments, the crevice holds water and may allow continued attack even when nearby surfaces are dry. In sea water, the severity of attack in stainless steel crevices depends on the ratio of the crevice area to the cathodic surface area outside the crevice. If the cathodic area is large relative to crevice area, corrosion is promoted.

Copper-containing alloys

Copper alloys often show only weak crevice corrosion. This is especially the case if the copper alloy is coupled to a less noble alloy such as steel. The corrosion of the steel is stimulated by the galvanic effect caused by the coupling of dissimilar metals. Hence, the sacrificial corrosion of the steel protects the copper alloy (Fig. 2.9). See Chap. 16, "Galvanic Corrosion."

Figure 2.9 Severely attacked crevice between rolled brass heat exchanger tube and mild steel tube sheet.

Solution-conductivity

Crevice corrosion is markedly increased as water conductivity increases.

Solution pH

Acidic pH helps break down protective oxides on stainless steels. Corrosion usually develops faster and is more severe as pH decreases. At very low pH, however, attack inside crevices may be no more severe than on regions outside the crevice.

Abrasion

If closely fitting surfaces rub against one another, protective oxides may be dislodged, stimulating attack (Figs. 2.10 and 2.11). This is especially true if temperature variation is great and if differences in thermal expansion of abutting metals are large. Often such attack may resemble fretting (see Glossary).

Galvanic effects

Coupling of alloys which vary widely in galvanic potential will stimulate attack. Often such coupling is not obvious. A copper-impregnated valve bushing caused severe corrosion of a nodular cast iron fitting (Figs. 2.12 and 2.13). Although a galvanic effect was suspected, crevice corrosion was widespread in this system even when dissimilar alloys were not in contact.

Figure 2.10   Wastage in a circumferential region where a 90:10 cupronickel main condenser tube passed through a baffle. In places, metal loss was greater than 25% of the nominal wall thickness.

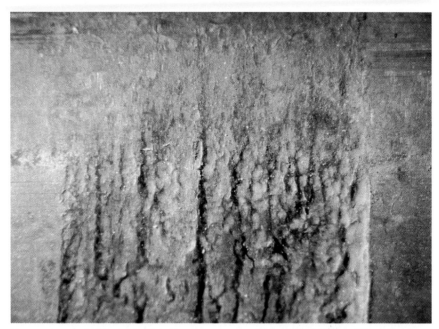

Figure 2.11 As in Fig. 2.10. Note the more severe attack at the bottom of the tube where the tube rested against its support. (Magnification: 7.5×.)

Figure 2.12 Mild steel valve block split in half with a plunger shaft still in place. Note the ragged ring etched into the shaft throat. See Fig. 2.13.

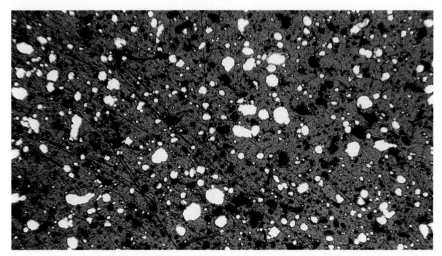

Figure 2.13 Small particles of copper embedded in shaft bushing material (shown in Fig. 2.12). (Magnification: 55× unetched.)

Identification

Wastage at the crevice is obvious. Corrosion follows the dimensions of abutting surfaces. Corrosion beneath washers and gaskets produces ghost images of the abutting surfaces (Figs. 2.14 and 2.15). At threaded joints, attack is confined to overlapping areas adjacent to

Figure 2.14 Close-up of annular regions below rubber O-rings on a cast iron valve block. Note how damage varies from hole to hole, probably due to variation in the crevice geometry.

Figure 2.15   As in Fig. 2.14. (Magnification: 7.5×.)

fluids (Fig. 2.16). In stainless steel joints, pitting may be pronounced (Fig. 2.17). General wastage almost always occurs at cast iron and lower alloy steel joints (Figs. 2.18 and 2.19).

Figure 2.16   Austenitic stainless steel pipe nipple attacked at threads by acidic chloride-containing solution. (Magnification: 7.5×.)

Figure 2.17 Pitting at a free-machining austenitic stainless steel threaded valve throat. The valve controlled flow of a chlorine-containing, low-pH biocide.

Corrosion products are almost always absent in stainless steel crevices. Areas just outside stainless crevices are stained brown and orange with oxides (Figs. 2.20 and 2.21). Metal ions migrate out of the crevice. Precipitation occurs by reactions similar to Reactions 2.3 and 2.4. Crevice interiors remain relatively free of rust (Figs. 2.16 and 2.17).

On mild steel and cast irons, rust accumulates at crevice mouths. Darker oxides often are present within crevices (Figs. 2.18 and 2.19).

Figure 2.18 Wasted threads on small-diameter mild steel pipe.

Figure 2.19   Gray cast iron elbow showing thread wastage.

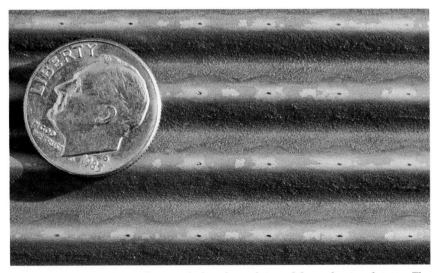

Figure 2.20   Austenitic stainless steel plate from plate-and-frame heat exchanger. The orange oxide was formed from corrosion product originating at the regularly spaced pits. Pits are present near points of contact between adjacent plates. Corrugations run at right angles on adjacent plates.

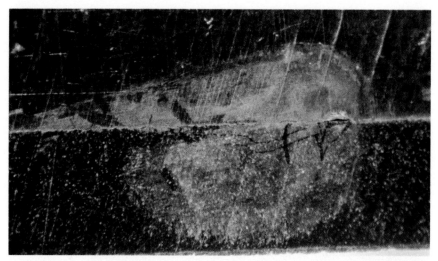

Figure 2.21   Orange oxides weeping from a tight seam in a martensitic (440) stainless steel surgical scissors. (Magnification: 15×.)

Chemical analysis of surfaces within crevices often reveals high concentrations of chloride. Chemical spot tests can indicate acidic pH.

Frequently, so-called crevice washers are used in coupon studies to test the environment for its ability to produce crevice corrosion (Fig. 2.22). There are several designs; most consist of a small Teflon* washer with radially oriented, wedge-shaped teeth. The washer is held to the coupon by a mounting bolt that passes through a central hole. The spaces between teeth form small crevice-shaped areas in which attack may occur (Fig. 2.23). The test is somewhat subjective and is not easily quantified. Using this test, attack in crevices either occurs or does not.

Elimination

Crevice attack depends on establishing a crevice geometry and allowing water to enter the crevice. Hence, forms of prevention may fall into one of three general classifications:

- Eliminate the crevice.
- Remove all moisture.
- Seal the crevice.

---

* Teflon is a registered trademark of E. I. du Pont de Nemours & Company.

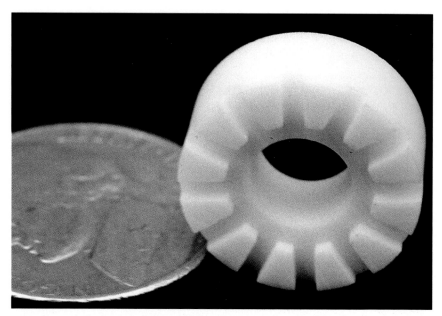

Figure 2.22   Crevice washer. The radially oriented, wedge-shaped washer teeth provide small areas between which crevices are formed.

Figure 2.23   Shallow pitting in crevice areas on a 304 stainless steel coupon exposed to a misting atmosphere. Note the relatively clean areas where the washer teeth contacted the coupon surface. (Magnification: 7.5×.)

Additionally, crevice corrosion can be reduced by two techniques used successfully on most aqueous corrosion—chemical inhibition and cathodic protection. However, both these techniques may be cost prohibitive.

The techniques that are most effective are usually system specific. In general, the following steps have been shown to be effective:

1. Do not use riveted joints.

2. Employ sound welding practice. Porosity should be minimized, and full penetration is required. Continuous welds are preferred to many short runs.

3. Allow for drainage of water.

4. Paint, grease, caulk, solder, or otherwise seal known crevices before exposure to water.

5. In critical applications, cap bolts with grease-filled containers. In some instances, epoxy materials have been used to fill such caps (with limited success).

6. Coat condenser water boxes with protective materials. Special attention should be paid to the tube sheet area and any sharp projections such as nuts and bolts.

7. Avoid using hydrochloric acid to clean stainless systems if any alternative is possible. (Chloride will concentrate in preexisting crevices during cleaning and may not be removed subsequently.)

8. In severe environments, welding tube ends into tube sheets may reduce crevice attack.

9. During prolonged outages, remove wet packing and clean gasketed surfaces thoroughly.

10. Judicious use of chemical inhibitors or cathodic protection may directly and indirectly reduce crevice attack. However, if conditions are very severe, economics may make such applications impractical.

11. Make sure all gaskets are in good repair and bolts are properly tightened.

12. Remove detritus, trash, or accumulations of foreign material on a regular basis.

13. In misting areas, exercise special care in maintaining paint layers. Regularly inspect known problem areas.

Cautions

Occasionally corrosive bacteria may colonize crevices. Although rare compared to ordinary crevice attack, microbiologically influenced attack should always be suspected if large amounts of sulfide or biological material are present in or near troubled crevices.

Related Problems

See Chap. 3, "Tuberculation"; Chap. 7, "Acid Corrosion"; and Chap. 16, "Galvanic Corrosion."

CASE HISTORY 2.1

Industry:	Steel—continuous caster
Specimen Location:	Stainless steel corrugated plates from a plate-and-frame heat exchanger. Closed chill water cooling loop used to cool continuous caster mold.
Specimen Orientation:	Vertical
Environment:	Cooling water side: Softened cooling water with chromate treatment
	Mold side: Mold cooling water after cooling molten steel
Time in Service:	Less than 1 year
Sample Specifications:	17 in. (43 cm) by 18 in. (46 cm) plates 0.023 in. (0.058 cm) thick, corrugated 316 stainless steel

Fine, short, cracklike pits run transversely across the apex of many corrugations on the stainless sheets (Figs. 2.24 and 2.25). The corrosion sites originate at the apex of the corrugations at small circular discolorations, spaced apart exactly the peak-to-peak distance between corrugations. All cracklike pits are aligned in the same direction, regardless of their locations on the plate.

Close examination of plate surfaces with a low-power stereomicroscope indicates that the attack sites are cracklike in appearance, but they are actually greatly elongated holes or tunnels that originate at pit sites at the apex of corrugations then propagate down through the plate wall (Fig. 2.26).

Pitting is caused by attack at crevices formed by closely overlapping corrugations of adjacent plates. Chloride concentration is high within pits. The tunneling of the pits and their precise alignment in the stainless steel plate rolling direction indicates that the tunneling is occurring along nonmetallic inclusions within the steel.

Figure 2.24   Cracklike corrosion site at apex of corrugation.

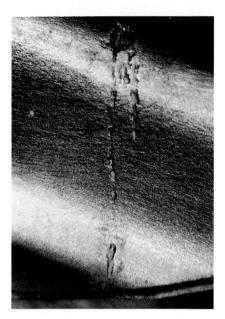

Figure 2.25   Pit-site and tunnel propagation down the side of the wall. (Magnification: 7.5×.)

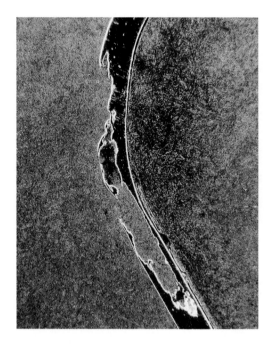

Figure 2.26   As in Fig. 2.25. Longitudinal cross section through tunnel. (Magnification: 15×.)

CASE HISTORY 2.2

Industry:	Chemical process
Specimen Location:	Heat exchanger-recycle compressor aftercooler in an ethylene unit
Specimen Orientation:	Horizontal
Environment:	Shell side: 70–90°F (21–32°C), total hardness ~425 ppm, pressure 60 psig (414 kPa), Ca ~275, P alk ~40, pH 8.5–9.0, M alk ~200, conductivity ~2000–3500 µmhos; phosphate treatment, dispersant, gaseous chlorine 0.3–0.2 ppm
	Tube side—ethylene gas: 190°F (88°C) inlet, pressure 6000 psig (41,400 kPa)
Time in Service:	14 years
Sample Specifications:	1.25 in. (3.18 cm) outer diameter, 0.160 in. (0.406 cm) average wall thickness, SA 210 steel (carbon steel) tube. Failure occurred near the middle of the exchanger.

An ethylene heat exchanger tube failed in service. The tube was severely thinned beneath a baffle and tore in half (Fig. 2.27).

Wastage was caused by long-term crevice corrosion. Attack was much more severe beneath the baffle than elsewhere. Subsequent investigation revealed severe damage at many baffles.

Figure 2.27   Close-up of severely thinned heat exchanger tube in a baffle.

Concentration of acidic deposits had occurred at crevices between tubes and baffles. Sufficient wastage eventually accumulated at the baffle crevice and caused catastrophic failure.

CASE HISTORY 2.3

Industry:	Primary metals
Specimen Location:	Rolling mill drive motor heat exchanger (air cooler)
Specimen Orientation:	Horizontal (tubes)
Environment:	Tube side: Mill supply water, clarified river water, 60–90°F (16–32°C)
	Shell side: Ambient air
Time in Service:	7 years
Sample Specifications:	68, 1 in. (2.54 cm) outer diameter, admiralty brass tubes rolled into mild steel tube sheets

A rolling mill drive motor failed in service. Investigation revealed the motor was short circuited by water in-leakage. This forced a prolonged shutdown, as the motor could not be replaced on site. High humidity was initially assumed to be the problem. Upon close inspection of the motor air coolers, however, it was determined that water leaked from headboxes into the air plenum spaces. Severe corrosion at tube rolls was observed (Fig. 2.9).

The tube sheet was wasted at the crevice between the sheet and the rolled end. In places, there was about ¼ in. (0.64 cm) clearance between the tube and tube sheet.

Wastage was caused by crevice corrosion, accelerated by the difference in tube and tube sheet metallurgies. The brass tube, being more noble, was cathodically protected by corrosion of the surrounding mild steel tube sheet. However, the galvanic effect was secondary to the primary cause of failure, namely, crevice corrosion.

Because of time and equipment constraints, the failed exchangers were repaired and remachined on face plates, and the crevices were sealed with phenolic materials as a stopgap measure. New exchangers were ordered.

CASE HISTORY 2.4

Industry:	Nuclear utility
Specimen Location:	Closed-cycle service water heat exchanger (turbine cooling)
Specimen Orientation:	Horizontal
Environment:	Tube side: Sea water, 0.25 ppm total residual chlorine, continuous, 40–90°F (4–32°C), pH 7.5–8.5, pressure estimated 15–50 psig (103–345 kPa), tubes supported by support plates
	Shell side: Demineralized water, sodium nitrite, 90–120°F (32–49°C), pressure 50–75 psig (345–517 kPa)
Time in Service:	12 years
Sample Specifications:	1 in. (2.54 cm) outer diameter, 0.050 in. (0.13 cm) wall thickness, 90:10 cupronickel tubes

As part of regular inspection procedures, several sections of tubing were removed. Severe thinning was found on external surfaces near tube supports (Figs. 2.10 and 2.11). Corroded areas had smoothly undulating interiors. Corrosion products and deposits were largely absent, as these sections were mechanically cleaned before inspection. However, visual reports suggested little material was present before cleaning.

In some areas, up to 25% of the wall thickness was gone at the grooved regions. More metal was lost from the bottoms of the tubes than from the tops (Fig. 2.11).

It would not be expected that crevice corrosion could easily occur in demineralized water. However, this exchanger was subject to extended outages with no lay-up procedures. Under such conditions, crevice attack is more likely.

3

Tuberculation

General Description

Tubercles are mounds of corrosion product and deposit that cap localized regions of metal loss. Tubercles can choke pipes, leading to diminished flow and increased pumping costs (Fig. 3.1). Tubercles form on steel and cast iron when surfaces are exposed to oxygenated waters. Soft waters with high bicarbonate alkalinity stimulate tubercle formation, as do high concentrations of sulfate, chloride, and other aggressive anions.

Tubercles are much more than amorphous lumps of corrosion product and deposit. They are highly structured. Structure and growth are interrelated in complex ways.

Incipient growth

In oxygenated water of near neutral pH and at or slightly above room temperature, hydrous ferric oxide [$Fe(OH)_3$] forms on steel and cast irons. Corrosion products are orange, red, or brown and are the major constituent of rust. This layer shields the underlying metal surface from oxygenated water, so oxygen concentration decreases beneath the rust layer.

More reduced forms of oxide are present beneath the rust layer. Hydrous ferrous oxide ($FeO \cdot nH_2O$), that is, ferrous hydroxide [$Fe(OH)_2$]

Figure 3.1   Utility service water system pipe almost plugged by heavy tuberculation.

is next to the metal surface. A black, magnetic hydrous ferrous ferrite layer ($Fe_3O_4 \cdot nH_2O$) can form between the ferric and ferrous oxides. These layers are shown schematically in Fig. 3.2. The topmost layer is orange and brown, whereas the underlying layers are usually black.

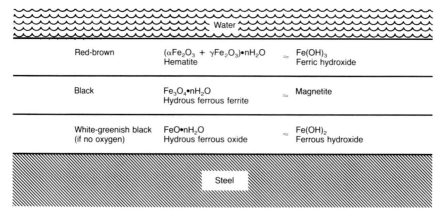

Figure 3.2   Incipient rust layer on steel in oxygenated water. (*Courtesy of National Association of Corrosion Engineers, Corrosion '91 Paper No. 84 by H. M. Herro.*)

As rust accumulates, oxygen migration is reduced through the corrosion product layer. Regions below the rust layer become oxygen depleted. An oxygen concentration cell then develops. Corrosion naturally becomes concentrated into small regions beneath the rust, and tubercles are born.

Structure

All tubercles have five structural features in common:

1. Outer crust
2. Inner shell
3. Core material
4. Fluid-filled cavity
5. Corroded floor

These features and their associated compounds are shown in Fig. 3.3. Typical reactions and their locations appear in Fig. 3.4.

Outer crust. A friable outer crust forms atop the tubercle. The crust is composed of ferric hydroxide (hematite), carbonates, silicates, other precipitates, settled particulate, and detritus. Ferrous ion and ferrous hydroxide generated within the tubercle diffuse outward through fis-

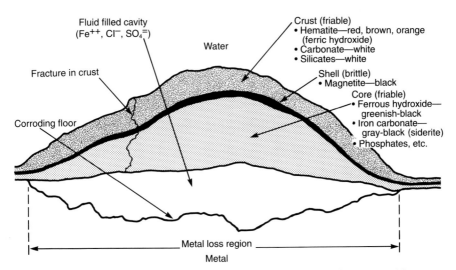

Figure 3.3 Well-developed tubercle shows chemical compounds and structure. (*Courtesy of National Association of Corrosion Engineers, Corrosion '91 Paper No. 84 by H. M. Herro.*)

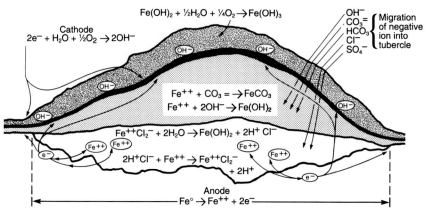

Figure 3.4 As in Fig. 3.3, shows chemical reactions within the tubercle. (*Courtesy of National Association of Corrosion Engineers, Corrosion '91 Paper No. 84 by H. M. Herro.*)

sures, where they encounter dissolved oxygen. Ferric hydroxide is produced and precipitates atop the tubercle as in Reaction 3.1:

In crust: $\qquad\qquad Fe(OH)_2 + \tfrac{1}{2}H_2O + \tfrac{1}{4}O_2 \rightarrow Fe(OH)_3$ $\qquad\qquad$ (3.1)

Inner shell. Just beneath the outer crust a brittle, black magnetite shell develops. The shell separates the region of high dissolved-oxygen concentration outside the tubercle from the very low dissolved-oxygen regions in the core and fluid-filled cavity below (Fig. 3.5). The shell is mostly magnetite and thus has high electrical conductivity. Electrons generated at the corroding floor are transferred to regions around the tubercle and to the shell, where cathodic reactions produce hydroxyl ion, locally increasing pH. Dissolved compounds with normal pH solubility, such as carbonate, deposit preferentially atop the shell where pH is elevated as in Reaction 3.2:

At shell: $\qquad\qquad e^- + H_2O + \tfrac{1}{2}O_2 \rightarrow 2OH^-$ $\qquad\qquad$ (3.2)

A profile of oxygen and pH concentration is shown in Fig. 3.5. A cross section of the outer crust, inner magnetite shell, and core material is shown in Fig. 3.6.

Core. Friable core material is present beneath the magnetite shell (Fig. 3.6). The core consists mostly of ferrous hydroxide formed by Reaction 3.3:

Core: $\qquad\qquad Fe^{++} + 2OH^- \rightarrow Fe(OH)_2$ $\qquad\qquad$ (3.3)

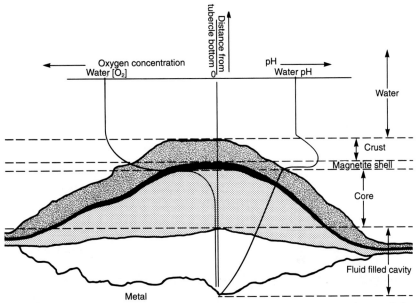

Figure 3.5 Schematic pH and oxygen concentration profiles in an active tubercle. Below the magnetite shell oxygen concentration decreased sharply. pH rises above the magnetite shell due to cathodic hydroxyl-ion generation and falls below the shell due to concentration of acidic anion. (*Courtesy of National Association of Corrosion Engineers, Corrosion '91 Paper No. 84 by H. M. Herro.*)

The hydroxyl ions migrate inward, attracted by the positive charge that is produced by the ferrous ion generated near the corroding surface (Fig. 3.4). Other anions such as carbonate, chloride, and sulfate also concentrate beneath the shell. Carbonate may react with ferrous ions to form siderite ($FeCO_3$) as in Reaction 3.4 (Fig. 3.7):

Core: $$Fe^{++} + CO_3^= \rightarrow FeCO_3 \qquad (3.4)$$

Other compounds, including phosphates, may be found within core material.

Cavity. A fluid-filled cavity is sometimes present beneath the core (Fig. 3.3). The cavity may be huge as in Fig. 3.8 or small as in Fig. 3.9. The cavity may result, in part, from acidic conditions internally. Acid conditions may prevent precipitation of oxides and hydroxides inside the tubercle.

Floor. A localized corroded region is always present beneath the tubercular mound. The depression is usually much broader than it is

Figure 3.6 Tubercle cross section shows black magnetite-rich shell beneath deposit and hematite cap. Note the core material below the magnetite shell. (Magnification: 7.5×.) (*Courtesy of National Association of Corrosion Engineers, Corrosion '89 Paper No. 197 by H. M. Herro.*)

deep, forming a shallow dish-shaped bowl (Fig. 3.10). Iron dissolves, forming ferrous ions according to Reaction 3.5:

$$\text{Floor:} \qquad \text{Fe} \rightarrow \text{Fe}^{++} + 2e^- \qquad\qquad (3.5)$$

If chlorides are present internally, acidity increases due to hydrolysis:

$$\text{Floor-core boundary:} \quad \text{Fe}^{++}2(\text{Cl}^-) + 2\text{H}_2\text{O} \rightarrow \text{Fe(OH)}_2\downarrow + 2\text{H}^+\text{Cl}^- \quad (3.6)$$

$$\text{Near floor:} \qquad 2\text{H}^+\text{Cl}^- + \text{Fe}^{++} \rightarrow \text{Fe}^{++}2(\text{Cl}^-) + 2\text{H}^+ \qquad (3.7)$$

Hence, acidity will become pronounced if a readily hydrolyzable anion such as chloride is present. Similarly, sulfate may accelerate attack by depressing internal pH.

Locations

Tubercles grow on nonstainless steels and some cast irons. Sensitized stainless steel and a few other alloys are rarely affected. Surfaces must

Figure 3.7 Siderite ($FeCO_3$) nodules inside a tubercle revealed by breaking open the magnetite shell. (*Courtesy of National Association of Corrosion Engineers, Corrosion '89 Paper No. 197 by H. M. Herro.*)

contact oxygenated water during growth and must remain wet for extended periods.

Common cooling water components that suffer tuberculation are listed below:

- Service water system piping
- Heat exchanger water boxes
- Steel heat exchanger tubes
- Pump components
- Storage tanks
- Cooling tower bolts, fittings, and sheet metal

Any uncoated or untreated steel or cast iron component may be attacked if it contacts oxygenated water for a prolonged period.

Deposits stimulate tubercle formation. Hence, regions where foreign material accumulates are common tubercle breeding grounds. Stagnant, low-flow areas also promote tubercle growth.

The relationship between flow and tubercle growth is dichotomous. Low flow may stimulate growth, but zero flow (so that water contacting

Figure 3.8   Large tubercle broken open to reveal the internal cavity. The cavity is filled with fluid in service.

surfaces contains no oxygen) stops attack. If flow is high, turbulence may dislodge incipient tubercles. Thus, pump impellers and other apparatuses experiencing severe turbulence almost never show tubercular growth unless they have been out of service for an extended period.

Critical Factors

Dissolved oxygen

The most important factor controlling tubercle growth is dissolved oxygen concentration. If dissolved oxygen concentration is very low, tubercular growth is severely retarded. However, oxygen-saturated waters are not required for growth, as near-stagnant systems often experience severe attack. In waters containing no dissolved oxygen, growth associated with oxygen concentration cell action ceases, since the driving force for tubercle growth is differential aeration.

Flow

Tubercles form under high- and low-flow conditions. Flow directly influences tubercle morphology. When flow is great, tubercles elongate

Figure 3.9 Small tubercle cross section. The steel surface is at the bottom, and the red material is an epoxy mounting medium. Note the small, shallow central cavity. (Magnification: 7.5×.)

in the direction of the flow (Fig. 3.11). Flow also affects growth by replenishing dissolved oxygen, aggressive anions, chemical inhibitors, and suspended particulate. If flow is very high, turbulence will dislodge tubercular structures.

Figure 3.10 Large-diameter steel pipe wall cross section. Note the very shallow, broad depression beneath the tubercle. (*Courtesy of National Association of Corrosion Engineers, Corrosion '91 Paper No. 84 by H. M. Herro.*)

Figure 3.11 Tubercles elongated by flow in a mill water supply line.

Biological interactions

The role organisms play in tubercle development is not clear. Tubercles often form in boiler superheaters during idle periods. Hence, it would appear microorganisms do not have to be present for tubercle formation. It is apparent that large numbers of microorganisms, and even macroorganisms, colonize tubercles. The degree to which such organisms influence tubercular development is uncertain (see Chap. 6, "Biologically Influenced Corrosion"). Sulfate-reducers and acid-producing bacteria probably accelerate attack.

Aggressive anion

Waters containing high concentrations of chloride, sulfate, and other aggressive anions stimulate tubercle growth. Very high concentrations of chloride and sulfate can be found internally in many fast-growing tubercles. Hydrolysis produces acidic conditions internally, thereby stimulating growth.

As bulk water pH falls, tubercle numbers and size tend to increase. At sufficiently low pH, however, precipitates and oxides cannot form (i.e., they are dissolved) and tubercular structures cannot exist.

Identification

The presence of tubercles is usually obvious. Friable brown and orange nodular encrustations on mild steel and cast iron cooling water components are almost always tubercles (Figs. 3.12 through 3.14). The presence of a crust, shell, core, cavity, and corroded floor are definitive (Fig. 3.3). Careful analysis can provide considerable information concerning growth, chemical composition, and associated metal loss.

Structure

When dry, tubercles are usually brittle and can be crushed by gentle pressure with a finger. Tubercle caps can be dislodged whole with a hard implement (Figs. 3.15 through 3.17). The physical strength of the tubercle is related to the thickness of the magnetite shell and densities of crust and core material. A typical magnetite shell in cross section is shown in Fig. 3.6. Dense, thick shells harden structures. These harder, denser tubercles have usually grown at a slower rate than lower-density tubercles. Thin magnetite shells are usually indicative of fast growth. Thus, thin magnetite shells are undesirable.

Multiple magnetite shells may form by successive fracture. Ferrous species spew out of the fractured shell and are quickly oxidized to form a new ferric hydroxide crust. Beneath the new crust, another mag-

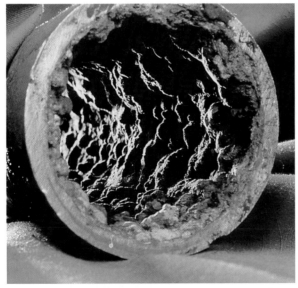

Figure 3.12   Heavily tuberculated 3-in. (7.6-cm) outer diameter steel mill water supply line.

Figure 3.13   Emergency supply line shows moderate tuberculation.

netite shell forms (Fig. 3.18). As tubercles age, their internal structure and outer morphology are altered.

Tubercles may become very large [up to 6-in. (15 cm) diameter structures have been reported]. The amount of metal loss is usually much less than the accumulated corrosion product and deposit might suggest. The average tubercle density when dry may be less than 1 g/cm^3. Hence, tubercles occupy much larger volumes than the metal loss they cap. Tubercle height may be from 5 to 30 times as great as the metal-loss depth below. The depth of metal loss below a tubercle can be roughly estimated using the above stated rule of thumb. Corrosion exceeding 50 mil per year locally is severe; average corrosion rates of about 10–20 mil per year beneath active tubercles are typical.

Chemical composition

The outer crust is composed of rust (hematite), precipitate, and settled particulate. Treatment chemicals may also deposit preferentially atop tubercles in response to associated corrosion. It is common to find several percent of zinc and phosphorus compounds in tubercles that grow in zinc- and phosphate-treated waters. Silicates also can be found in

Figure 3.14   Small, hard tubercles on an essential service water system pipe in a nuclear utility.

Figure 3.15   Knife dislodging a dry tubercle from a steel pipe surface.

Figure 3.16 As in Fig. 3.15 but tubercle cap is completely dislodged.

conjunction with associated treatments. High concentrations of carbonate may be found in the crust. Effervescence upon exposure to a few drops of acid usually proves the presence of carbonate.

Table 3.1 gives the local elemental composition of three different tubercles from three different systems formed under different chemical treatments. At the floor of each tubercle, the concentration of chlorine and sulfur is higher than in the crust. The concentration of most crust elements, except that of iron, also decreases near the tubercle floor. The crust contains traces of treatment chemicals including zinc, phosphorus, and silicon. Tubercle 1 contains up to 40% silicon in the crust, which strongly suggests accumulation of silt by settling of particulate.

Figure 3.17 Intact tubercle showing physical integrity after removal from surface in Figs. 3.15 and 3.16.

Figure 3.18 Multiple magnetite shells in a small tubercle. Multiple shells form due to successive fracture during growth. (Magnification: 2×.)

TABLE 3.1   Compositions of Three Tubercles at Crust and Floor

(Tubercle compositions were measured using energy dispersive x-ray analysis)*

Element	Tubercle 1 (phosphate treatment)		Tubercle 2 (silicate treatment)		Tubercle 3 (phosphate zinc treatment)	
	Crust	Floor	Crust	Floor	Crust	Floor
Fe	30	86	90	95	66	70
Al	14	—	4	1	—	—
Si	40	—	5	1	6	2
S	1	2	1	3	5	19
Cl	—	11	—	—	—	2
Mg	1	—	—	—	3	—
Ca	3	—	—	—	4	—
P	2	—	—	—	7	2
Zn	—	—	—	—	5	—
Ti	1	—	—	—	—	—
Mn	4	1	—	—	3	<1
K	4	—	—	—	—	—
Cu	—	—	—	—	1	2
Na	—	—	—	—	—	2

* All concentrations are expressed as percent element.

As a result of the concentration of acidic species, such as chloride and sulfate, material scraped from the inside of tubercles is virtually always acidic when mixed with water. Acidity varies not only from tubercle to tubercle but also from place to place in a given tubercle. Acidity is greatest near the corroded metal surface. The size of the fluid-filled cavity can indicate acidity. The larger the cavity, the more acidic the internal environment.

When first exposed to air, tubercles are often quite dark both externally and internally (Figs. 3.19 and 3.20). After prolonged exposure to air, however, tubercle color changes to red, brown, and orange (Figs. 3.21 and 3.22). The more reduced ferrous hydroxide is converted to ferric hydroxide. A very rapid conversion to lighter colors sometimes signifies high tubercle acidity.

Corroded floor

The corroded tubercle floor is almost always a dish-shaped depression, much wider than it is deep (Fig. 3.23). Undercutting is very rare. The metal-loss width almost exactly matches the tubercular mound width. Corrosion rates exceeding 50 mil per year are rare, except when tubercles are young. Average local corrosion rates are usually 20 mil per year or less.

If pH is unusually low within the tubercle, the floor may be heavily striated. Small, shallow parallel grooves will appear in the depressions beneath each tubercle (Fig. 3.24). The striations are caused by preferential corrosion along microstructural defects such as deformed metal,

Figure 3.19   Tubercle only a few minutes after removal from a water-filled pipe.

Figure 3.20   As in Fig. 3.19, with tubercle opened to show the dark interior.

Figure 3.21   As in Figs. 3.19 and 3.20, after several hours of exposure to air. Note the reddening.

Figure 3.22   As in Fig. 3.21, but opened to show color change of interior.

Figure 3.23   Perforation at a dish-shaped depression on the internal surface of a large-diameter steel pipe. A large tubercle capped the depression but was dislodged during tube sectioning. (*Courtesy of National Association of Corrosion Engineers, Corrosion '89 Paper No. 197 by H. M. Herro.*)

Figure 3.24 Striated, corroded depressions are revealed on a large-diameter pipe after tubercles were removed. Striated surfaces are caused by preferential corrosion along microstructural irregularities in the steel. Such attack can be caused by low pH conditions.

chemical inhomogenities, and second-phase material. The defects are elongated and frozen into stringlike artifacts, paralleling the original rolling direction of the steel.

The corroded floor usually is covered with porous friable corrosion product containing ferrous hydroxide, which may form in place by conversion of the steel surface. If acidity is significant, the thickness of this corrosion product layer is slight.

Elimination

There are three ways in which tubercles may be prevented from forming and, once formed, from growing: chemical treatment, altering system operation, and material substitution.

Chemical treatment

Any chemical treatment that reduces general corrosion rates associated with oxygen corrosion will decrease tuberculation. The exact treatment that is best is system dependent. Water chemistries and operating practices may differ widely even among similar industries using similar equipment. Methods employing chemical inhibitors and dispersants are typical in cooling water systems.

Chemical inhibitors, when added in small amounts, reduce corrosion by affecting cathodic and/or anodic processes. A wide variety of treatments may be used, including soluble hydroxides, chromates, phosphates, silicates, carbonates, zinc salts, molybdates, nitrates, and magnesium salts. The exact amount of inhibitor to be used, once again, depends on system parameters such as temperature, flow, water chemistry, and metal composition. For these reasons, experts in water treatment acknowledge that treatment should be "fine tuned" for a given system.

No matter what chemical treatment is used, it will almost always work better on a comparatively clean metal surface. Tubercles, with rare exceptions, grow more rapidly when young than when old. There are assertions that tubercles go through dormant periods when growth is slow, and there is some evidence that growth is slow when tubercles are very old. Nevertheless, experience (sometimes painful) has shown that removal of tubercular encrustations is beneficial when chemically treating a system. There are numerous methods of removal, including water blasting, mechanical abrasion, air rumbling and, most effective of all, chemical cleaning.

The use of dispersants is highly recommended in systems containing silt, sand, oil, grease, biological material, and/or other foreign material. Not only does increased dispersion generally increase the effectiveness of chemical inhibition, it also prevents nucleation of oxygen concentration cells beneath foulants.

Altering system operation

Because of the diversity of cooling water systems, only general guidelines for altering system operation can be given.

Flow is very important in controlling tuberculation. If water flow is zero, dissolved oxygen concentration in regions far from air-liquid boundaries would eventually fall to zero and oxygen concentration cell activity would cease. However, it is *not* recommended that flow be suspended in any cooling water system, since attendant corrosion, fouling, and biological activity may make tuberculation problems seem tame by comparison. Low flow (less than 3 ft/s [1 m/s]) can produce serious tubercular growth in a short time. It is usually best to maintain a flow of at least several feet per second to allow treatment chemicals to act effectively. A greater flow also retards fouling and thus reduces tubercular initiation sites.

For each rise in temperature of about 68°F (20°C), there is a corresponding doubling of corrosion rates. At about 160°F (71°C), oxygen corrosion on steel is at maximum. Above these temperatures, the solubility of oxygen falls to very low concentrations and tuberculation effectively ceases.

Cathodic protection using sacrificial anodes or applied current can retard or eliminate tuberculation. However, costs can be high and technical installation can be very difficult. Costs are markedly reduced if surfaces are coated (see Material substitution below).

Material substitution

Certain conditions, ultimately dictated by economics, make the substitution of more resistant materials a wise choice. Stainless steels (not sensitized) of any grade or composition do not form tubercles in oxygenated water; neither do brasses, cupronickels, titanium, or aluminum. However, each of these alloys may suffer other problems that would preclude their use in a specific environment.

A frequently useful and more economical solution to tuberculation and general rusting involves the use of protective coatings. Galvanized sheet steel is immune to tubercular growth until the zinc-rich layer is consumed. Epoxy and other field- and factory-applied coatings have been used extensively in cooling water systems to retard attack (Fig. 3.25). Most utility and larger condenser water boxes are coated with a protective layer. Application skill is critical to good service. Sharp corners, threads, edges, crevices, and projections are troublesome areas.

Figure 3.25   Epoxy-coated section of a tube sheet and rolled tube.

Cautions

Tubercles grow at about the same rate in large- and small-diameter piping. Hence, small-diameter pipes are likely to plug with tubercular material before perforation. The reverse is true of large-diameter pipes. Hence, replacement and cleaning of plugged small-diameter pipe may be necessary, but large-diameter pipes should not be ignored in a troubled system. Eventually, large-diameter pipes will perforate with ever-increasing frequency, even if flow has not been seriously diminished.

Related Problems

See Chap. 5, "Oxygen Corrosion"; Chap. 6, "Biologically Influenced Corrosion"; and Chap. 17, "Graphitic Corrosion."

CASE HISTORY 3.1

Industry:	Steel
Specimen Location:	Hot strip mill water supply main
Specimen Orientation:	Horizontal
Environment:	Lake Michigan water treated with dispersant; heavily fouled with grease and oil
Time in Service:	23 years
Sample Specifications:	54 in. (137 cm) outer diameter, 0.330 in. (0.840 cm), thick mild steel

After almost 20 years of service, many leaks developed throughout the mill water supply system. Upon inspection, it was found that perforations had occurred beneath large tubercles. Some tubercles were up to 6 in. (15 cm) in diameter. Accessible pipe was water blasted to remove tubercles, and the deepest metal-loss areas (several thousand) were weld repaired. Perforations ceased.

After 5 more years of service, a few leaks began to develop. Window sections cut from the main contained numerous tubercular encrustations (Fig. 3.26).

The entire internal surface was covered with a layer of dark greasy deposit and slime intermixed with reddish-brown oxides (oxides were dark immediately upon removal from systems) (Figs. 3.19 and 3.20). Analysis showed tubercular contents were fairly acidic.

Beneath each tubercle was a dish-shaped depression. Some depressions were as deep as half the intact wall thickness. Many shallow striation-streaked depressions were present on internal surfaces (Fig. 3.24). Circumferential welds were preferentially attacked (Fig. 3.27).

Recommendations were made to begin treatment with corrosion inhibitors and to make system operation changes to reduce grease and oil fouling. Other water chemistry recommendations involved reducing the amount of aggressive anion in solution and pursuing biocidal treatment.

Figure 3.26   Typical internal surface of heavily tuberculated mill water supply line about 16 hours after removal from system. Surface resembled Figs. 3.19 and 3.20 immediately after removal. Note the weld where corrosion product and deposit were removed.

Figure 3.27   As in Fig. 3.26, showing preferential weld attack.

CASE HISTORY 3.2

Industry:	Utility
Specimen Location:	Steel baffle plate from a test water box containing flowing cooling water
Specimen Orientation:	Vertical
Environment:	Tower cooling water, ~80°F (27°C), flow 3 ft/s (1 m/s), water, chlorides, and sulfates >2000 ppm, pH ~7.2–7.6
Time in Service:	2 weeks
Sample Specifications:	4⅞ in. by 1¾ in. by ⅛ in. (12.4 cm by 4.4 cm by 0.32 cm), 1010 steel, cold rolled

A test water box was installed during a 2-week trial to monitor corrosion and fouling in a utility cooling water system. A baffle plate from the test box was removed after the test. Small, hollow incipient tubercles dotted surfaces (Fig. 3.28). Small amounts of carbonate were present atop and around each tubercle. Each tubercle capped a small depression no deeper than 0.005 in. (0.013 cm) (Fig. 3.29). This indicated local average corrosion rates were as high as 130 mil/y (3.3 mm/y).

Each tubercle exhibited small clam-shell marks or growth rings (Fig. 3.30). Each ring was formed by fracture at the tubercle base during growth. Ejected internal contents rapidly deposited when contacting oxygenated waters. Tubercles were hollow (Fig. 3.31). Surfaces below the cap contained concentrations exceeding 10% of chloride and sulfate, producing severe localized acidic conditions.

Figure 3.28   Many incipient tubercles formed on a steel baffle plate in 2 weeks. (Magnification: 7.5×.) (*Courtesy of National Association of Corrosion Engineers, Corrosion '89 Paper No. 197 by H. M. Herro.*)

Figure 3.29   As in Fig. 3.28, but with tubercles removed to show pitlike depressions beneath each mound. (Magnification: 7.5×.) (*Courtesy of National Association of Corrosion Engineers, Corrosion '89 Paper No. 197 by H. M. Herro.*)

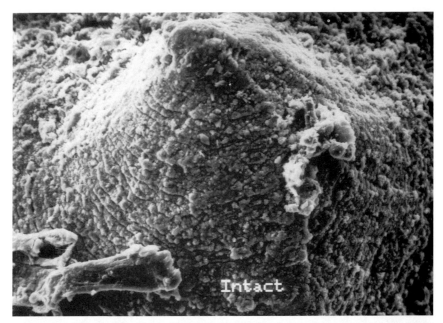

Figure 3.30   Scanning electron micrograph of tubercles in Fig. 3.28. Note the clam-shell growth steps formed by successive fractures at the tubercle base. Tubercle is about 200 μm in diameter. (*Courtesy of National Association of Corrosion Engineers, Corrosion '91 Paper No. 84 by H. M. Herro.*)

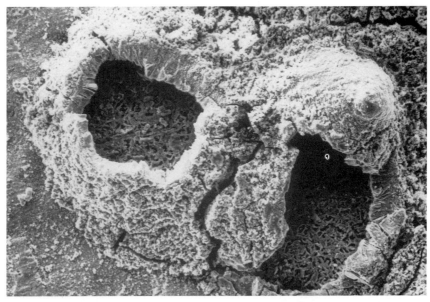

Figure 3.31   Tubercles in Fig. 3.30 broken open to reveal hollow interiors. (*Courtesy of National Association of Corrosion Engineers, Corrosion '91 Paper No. 84 by H. M. Herro.*)

CASE HISTORY 3.3

Industry:	Primary metal fabrication
Specimen Location:	Submerged pump housing
Specimen Orientation:	Vertical
Environment:	Pump housing submerged in a cooling water tank, pH 5–8, 55–90°F (13–32°C)
Time in Service:	More than 5 years
Sample Specifications:	Gray cast iron

The external surface of the pump housing was covered with large tubercles up to 2 in. (5 cm) in diameter (Fig. 3.32). Tubercles had hard, black, stratified shells, enveloping reddish-brown internal material. Surfaces beneath tubercles were corroded to a depth of as much as ¼ in. (0.54 cm). Regions below the tubercles were graphitically corroded (see Chap. 17).

Cooling water was used to cool hot aluminum. Calcium carbonate spotting had occurred at alkaline pH. After reducing pH, the staining problem disappeared, but corrosion increased substantially.

New chemical treatment programs were proposed that would permit reduction of staining at a higher pH and markedly reduce corrosion.

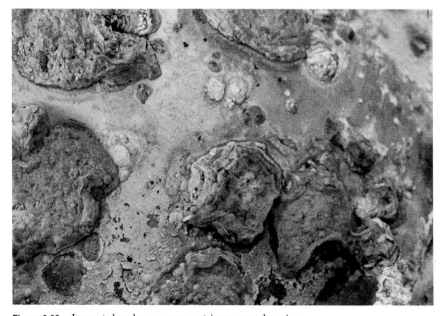

Figure 3.32   Large tubercles on gray cast iron pump housing.

CASE HISTORY 3.4

Industry:	Nuclear utility
Specimen Location:	Emergency service water pipe
Specimen Orientation:	Horizontal
Environment:	Near the end of a dead leg. Dispersant, tolyltriazole, silicate 15–25 ppm residual, bromine/chlorine, 1 hr/day
Time in Service:	7 years
Sample Specifications:	3½ in. (8.9 cm) outer diameter, mild steel pipe

An emergency system water pipe section was removed because of reduced flow. In places, as much as 40% of the internal cross-sectional area was blocked by tubercular growth (Fig. 3.1).

Tubercles consisted of hard, black oxide shells overlaid with friable carbonate-containing deposits. In places, several laminate black magnetite shells existed. The outer crust could be crushed by gentle pressure with a finger. Tubercles were riddled with white crystalline fibers. Other detritus was incorporated into the tubercle core and crust. Metal loss was less than 0.030 in. (0.076 cm) below each tubercle. Wall thickness was almost 0.25 in. (0.64 cm).

Chemical cleaning was suggested. A tannin-citric acid process was tested and produced good results.

Water chemistry and operating changes were suggested but were deemed inadvisable because of legal restrictions concerning discharge of chemicals into local waters.

CASE HISTORY 3.5

Industry:	Nuclear utility (1100 MW)
Specimen Location:	Turbine cooling water system piping, right angle L
Specimen Orientation:	Horizontal and vertical
Environment:	Water from cooling tower pump suction, pH 8.6–8.8, pressure 20–30 psi (140–210 kPa), flow 2–5 ft/s (0.2 to 1.5 m/s). Dispersant, 1–3 ppm tolyltriazole, sodium hypochlorite 2 hr/day to 0.8 free residual chlorine; 0.6–0.8 ppm total zinc and 0.1–0.2 ppm soluble zinc. Free chlorine maintained at 1 ppm for 5 consecutive days/month during the summer. Chemical treatment started after 2 years of no treatment. Water conductivity ~612 (μmhos/cm), turbidity 27 NTU (nephelometric turbidity units), chloride 110 ppm, sulfate 50 ppm, carbonate alkalinity ($CaCO_3$) 27 ppm, bicarbonate alkalinity ($CaCO_3$) 118 ppm
Time in Service:	3 years
Sample Specifications:	L-shaped pipe section of 2 in. (5 cm) carbon steel pipe socket welded to a 90° carbon steel elbow

Internal surfaces were moderately tuberculated (Fig. 3.14). Extremely thick, hard magnetite shells capped large internal cavities (Fig. 3.9). Pipe cross-sectional area was reduced by at least 30% in some places. Tubercles were aligned with flow, indicating that growth occurred during service. No failure occurred, and deepest metal loss was only 0.093 in. (0.033 cm) from the nominal pipe wall thickness of 0.225 in. (0.572 cm).

X-ray analysis of material scraped from internal surfaces indicated that it was 88% iron, 7% silicon, and 1% each of magnesium, aluminum, chlorine and sulfur.

This system was not treated except for chlorination during its first 2 years of existence. When pipes were inspected after 3 years, the amount and kind of corrosion appeared to have changed during 1 year of vigorous treatment. The 3-year inspection revealed that the tubercles had become appreciably darker and harder. This strongly suggested that more reducing conditions were established during the more recent treatment and that tubercle growth had slowed appreciably.

Chemical cleaning was considered since small-diameter pipe had been plugged in dead-leg regions. At the time of writing, no decision had been made concerning cleaning.

Underdeposit Corrosion

General Description

Deposits cause corrosion both directly and indirectly. If deposits contain corrosive substances, attack is direct; interaction with the aggressive deposit causes wastage. Shielding of surfaces below deposits produces indirect attack; corrosion occurs as a consequence of surface shielding provided by the deposit. Both direct and indirect attack may involve concentration cell corrosion, but indirect attack almost always involves this form of corrosion.

Concentration cell corrosion is caused by chemical concentration differences between areas contacting metal surfaces. An electrochemical cell is established by virtue of the difference between chemical compositions in the water on, and adjacent to, corroding areas.

Perhaps the best-known concentration cell, the oxygen concentration cell, involves differential aeration (Fig. 4.1A and B). Dissolved oxygen concentration inevitably decreases beneath deposits near actively corroding steel surfaces; corrosion beneath the deposit consumes oxygen. The deposit retards oxygen diffusion to regions near the corroding surface. Hence, an oxygen cell is established. As general oxygen corrosion proceeds, accumulation of corrosion products may also cause oxygen concentration cell formation. Oxygen concentration cells may also be found at crevices (see Chap. 2, "Crevice Corrosion"), beneath tuber-

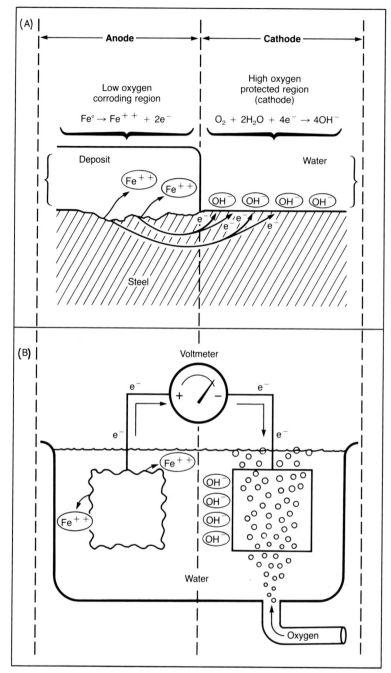

Figure 4.1 (A) Oxygen concentration cell corrosion beneath a deposit; (B) Oxygen concentration cell corrosion in a beaker containing an aerated piece of steel and an unaerated piece of steel. A potential develops between the aerated and unaerated steel pieces. Steel exposed to the lower dissolved oxygen concentration corrodes. Beneath a deposit (Fig. 4.1A) the oxygen-poor environment causes wastage.

cles (see Chap. 3, "Tuberculation"), or in any region partially shielded from oxygenated waters.

Severe concentration cell corrosion involves segregation of aggressive anions beneath deposits. Concentrations of sulfate and chloride, in particular, are deleterious. Acid conditions may be established beneath deposits as aggressive anions segregate to these shielded regions. Mineral acids, such as hydrochloric and sulfuric, form by hydrolysis. The mechanism of acid formation is discussed in Chap. 2.

Frequently, differential aeration and the concentration of aggressive ions beneath deposits combine to produce severe localized wastage on stainless steel (Figs. 4.2 and 4.3) and other passivating alloys such as aluminum, titanium, and nickel. However, differential aeration alone is not severe enough in most cases to initiate attack on stainless steels. Yet, in the presence of so-called depassivating agents, such as chloride ions, severe pitting may occur on stainless steels.

Pitting is also promoted by low pH. Thus, acidic deposits contribute to attack on stainless steels. Amphoteric alloys such as aluminum are harmed by both acidic and alkaline deposits (Fig. 4.4). Other passive metals (those that form protective corrosion product layers spontaneously) are similarly affected.

Indirect attack can also occur because of turbulence associated with flow over and around a deposit. Increased turbulence may initiate attack (see Chap. 11, "Erosion-Corrosion" and Chap. 12, "Cavitation Damage").

Locations

Attack always occurs beneath a deposit. Cooling water system deposits are ubiquitous. Deposits can be generated internally as precipitates, laid down as transported corrosion products, or brought into the system from external sources. Hence, underdeposit corrosion can be found in virtually any cooling water system at any location. Especially troubled

Figure 4.2 Severe localized wastage on a 316 stainless steel heat exchanger tube. Attack occurred beneath deposits, which were removed to show wastage.

Figure 4.3   A 304 stainless steel plate from a storage tank floor, which pitted beneath a deposit layer.

systems contain large amounts of silt, sand, grease, oil, biomass, precipitates, transported corrosion products, and other detritus. Deposits such as metal sulfides, manganese dioxide, and water-permeable materials tend to produce severe attack. Heat exchangers in which these materials are present are often troubled severely. Biological accumula-

Figure 4.4   Corrosion product mounds covering localized areas of metal loss on an aluminum heat exchanger tube. Attack initiated beneath a thin deposit layer.

tions such as slime layers are harmful. Attack is discussed in Chap. 6, "Biologically Influenced Corrosion."

Equipment in which water flow is slow or intermittent is subject to deposition and associated corrosion. Hence, service water-system components that operate intermittently frequently suffer attack. Deposits accumulate in narrow orifices, screens, long horizontal pipe runs, sumps, or at regions of constricted flow.

Components in which water temperature changes abruptly with distance, such as heat exchangers, tend to accumulate precipitates. Heater surfaces also accumulate precipitates if the dissolved species have inverse temperature solubilities. Systems in which pH excursions are frequent may accumulate deposits due to precipitation processes. Plenum regions, such as heat exchanger headboxes, tend to collect deposits.

Critical Factors

The most harmful deposits are those that are water permeable. Truly water-impermeable material is protective, since without water contacting metal surfaces corrosion cannot occur. Innately acidic or alkaline deposits are troublesome on amphoteric alloys (those attacked at high and low pH—e.g., aluminum and zinc).

Deposits containing carbonate can be protective. The carbonate buffers acidity caused by the segregation of potentially acidic anions in and beneath deposits. However, deposits are rarely composed of only a single chemical; mixed deposits are the rule. Deposit morphology also influences attack. Hence, although sometimes carbonate deposits are beneficial, they may also be deleterious.

Clean surfaces are almost always easier to protect chemically from corrosion. Chemical inhibition and inhibitors are often tested on relatively clean surfaces, for a variety of reasons. The effectiveness of almost all commonly used corrosion inhibitors increases as surface cleanliness improves.

Identification

Almost all cooling water system deposits are waterborne. It would be impossible to list each deposit specifically, but general categorization is possible. Deposits are precipitates, transported particulate, biological materials, and a variety of contaminants such as grease, oil, process chemicals, and silt. Associated corrosion is fundamentally related to whether deposits are innately aggressive or simply serve as an occluding medium beneath which concentration cells develop. An American

Society for Testing and Materials (ASTM) list of water-formed deposits is given in Table 4.1.

Oxides of manganese and iron are often found deposited together. Similar conditions cause oxidation of both iron and manganese ions. Exposure to oxygenated water, chlorination, and some microbiological processes causes such oxidation. Often, a few percent chlorine is found in deposits, possibly because of associated chlorination.

Manganese-rich deposits usually take one of three forms: A loosely adherent, friable, brown, or black deposit may occur (Fig. 4.5). A thin, dark, brittle, glassy manganese layer sometimes forms on heat transfer surfaces (Fig. 4.6). Nodular manganese deposits also occur (Fig. 4.7). Both nodular and glassy layers tend to occur on copper alloys.

Localized attack is common beneath manganese deposits on copper alloys (Fig. 4.8); it is most common beneath nodular and glassy deposits (Figs. 4.6 and 4.7). The exact corrosion mechanism or mechanisms responsible for pitting beneath manganese deposits on copper alloys are less than clear; no single explanation is consistent with all observed phenomena. Pits are usually filled with friable corrosion products. Pits associated with glassy manganese layers often contain little or no corrosion product and occur at breaks in the glassy layer. General wastage is common beneath porous, voluminous iron and iron-manganese deposits. Surfaces have irregular rolling contours similar to those found on surfaces corroded by oxygen (on steel and cast iron).

On galvanized steel, tubercles may develop rapidly at breaches in the zinc layer. Attack is frequently highly localized if aggressive ions such as chloride or sulfate concentrate beneath deposits (Fig. 4.9).

Stainless steels may pit or generally corrode beneath deposits (Figs. 4.10A and B through 4.12). Pitting on stainless steels results from a metal surface condition midway between complete passivity and general corrosion. As conditions beneath the deposit become more aggressive due to oxygen depletion, decreasing pH, and/or the concentration of depassivating agents, pits develop (Fig. 4.10A and B). Attack beneath deposits is unlikely unless depassivating agents and/or low pH are present. Attack is intense and localized; pronounced undercutting of pits is present if acid conditions are established below the deposit. Usually corrosion products are reddish brown and may surround the pits (Fig. 4.13). If deposits are sufficiently aggressive, corrosion may become general (Fig. 4.11). Borders separating corroded from unattacked regions are usually sharp, mirroring deposit patterns (Fig. 4.12). Corrosion products resemble rust formed on steel and cast iron surfaces but are considerably less voluminous; tuberculation is rare but can occur (e.g., at sensitized welds).

Contaminants

Silt, sand, concrete chips, shells, and so on, foul many cooling water systems. These siliceous materials produce indirect attack by establishing oxygen concentration cells. Attack is usually general on steel, cast iron, and most copper alloys. Localized attack is almost always confined to strongly passivating metals such as stainless steels and aluminum alloys.

Below silt accumulations on copper, brass, and cupronickel heat exchangers, a layer of bluish-white copper carbonate often forms (Figs. 4.14 and 4.15).

Precipitates

A wide variety of precipitates form in cooling water systems; carbonates, silicates, sulfates, and phosphates are common. Below and slightly above 212°F (100°C), calcite, aragonite, gypsum, hydroxyapatite, magnesium phosphate, anhydrite, and serpentine are commonly encountered (see Table 4.1).

Calcium carbonate makes up the largest amount of deposit in many cooling water systems (Fig. 4.16) and can be easily detected by effervescence when exposed to acid. Deposits are usually heavily stratified, reflecting changes in water chemistry, heat transfer, and flow. Corrosion may be slight beneath heavy accumulations of fairly pure calcium carbonate, as such layers can inhibit some forms of corrosion. When nearly pure, calcium carbonate is white. However, calcium carbonates are often intermixed with silt, metal oxides, and precipitates, leading to severe underdeposit attack.

Calcium carbonate has normal pH and inverse temperature solubilities. Hence, such deposits readily form as pH and water temperature rise. Copper carbonate can form beneath deposit accumulations, producing a friable bluish-white corrosion product (Fig. 4.17). Beneath the carbonate, sparkling, ruby-red cuprous oxide crystals will often be found on copper alloys (Fig. 4.18). The cuprous oxide is friable, as these crystals are small and do not readily cling to one another or other surfaces (Fig. 4.19). If chloride concentrations are high, a white copper chloride corrosion product may be present beneath the cuprous oxide layer. However, experience shows that copper chloride accumulation is usually slight relative to other corrosion product masses in most natural waters.

Sulfur compounds

Sulfides are directly aggressive to many metals. In particular, copper alloy corrosion is severe (Fig. 4.20). Sulfides are easily detected in the

TABLE 4.1 ASTM List of Water-Formed Compounds Found in Water-Formed Deposits

Compound	Formula	Compound	Formula
Acmite	$Na_2O \cdot Fe_2O_3 \cdot 4SiO_2$	Hydroxyapatite	$Ca_{10}(OH)_2(PO_4)_6$
Ammonium bicarbonate	NH_4HCO_3	Hydroczincite	$2ZnCO_3 \cdot 3Zn(OH)_2$
Analcite	$Na_2O \cdot Al_2O_3 \cdot 4SiO_2 \cdot 2H_2O$	Iron	Fe
Anhydrite	$CaSO_4$	Lazurite	$3Na_2O \cdot 3Al_2O_3 \cdot 6SiO_2 \cdot 2Na_2S$
Ankerite	$CaCO_3(Fe,Mg,Mn)CO_3$	Lead	Pb
Aragonite	$CaCO_3$	Lepidocrocite	$Fe_2O_3 \cdot H_2O$
Atacamite	$CuCl_4 \cdot 3Cu(OH)_4$	Libethenite	$4CuO \cdot P_2O_5 \cdot H_2O$
Azurite	$2CuCO_3 \cdot Cu(OH)_2$	Maghemite	Fe_2O_3
Barite	$BaSO_4$	Magnesia	MgO
Bayerite	$Al_2O_3 \cdot 3H_2O$	Magnesium chloride hydrate(basic)	$MgCl_2 \cdot 5Mg(OH)_2 \cdot 8H_2O$
Bloedite	$Na_2SO_4 \cdot MgSO_4 \cdot H_2O$	Magnesium hydroxyphosphate	$3Mg_3(PO_4)_2 \cdot Mg(OH)_2$
Boehmite	$Al_2O_3 \cdot H_2O$	Magnesite	$MgCO_3$
Bornite	Cu_5FeS_4	Magnetite	Fe_2O_3
Brochantite	$CuSO_4 \cdot 3Cu(OH)_2$	Malachite	$CuCO_3 \cdot Cu(OH)_2$
Brucite	$Mg(OH)_4$	Meta halloysite	$Al_2O_3 \cdot 2SiO_2 \cdot xH_2O$
Bunsenite	NiO	Meta thenardite	Na_2SO_4I
Burkeite	$Na_2CO_3 \cdot 2Na_2SO_4$	Montmorillonite	$Al_2O_3 \cdot 4SiO_2 \cdot H_2O \cdot nH_2O$
Calcium oxalate monohydrate	$CaC_2O_4 \cdot H_2O$	Mullite	$3Al_2O_3 \cdot 2SiO_2$
Calcium phosphate (dibasic)	$CaHPO_4$	Muscovite	$KAl_2(Si_3Al)O_{10}(OH_2F)_2$
Calcium pyrophosphate	$Ca_2P_2O_7$	Natrolite	$Na_2O \cdot Al_2O_3 \cdot 3SiO_2 \cdot 2H_2O$
Calcium sodium phosphate	$CaNaPO_4$	Nontronite	$H_4(Al,Fe)_2Si_2O_9$
Calcium sulfate	$CaSO_4$	Noselite	$4Na_2O \cdot 3Al_2O_3 \cdot 6SiO_2 \cdot SO_3$
Calcite	$CaCO_3$	Oldhamite	CaS
Calcium aluminate	$3CaO \cdot Al_2O_3 \cdot 6H_2O$	Olivine	$2(Mg,Fe)O \cdot SiO_2$
Cancrinite	$4Na_2O \cdot CaO \cdot 4Al_2O_3 \cdot 2CO_2 \cdot 9SiO_2 \cdot 3H_2O$	Para sepiolite	$2MgO \cdot 3SiO_2 \cdot 2H_2O$
Celestite	$SrSO_4$	Paratacamite	$CuCl_2 \cdot 3Cu(OH)_2$
Cementite	Fe_3C	Pectolite	$Na_2O \cdot 4CaO \cdot 6SiO_2 \cdot H_2O$
Cerrucite	$PbCO_3$	Periclase	MgO
α Chalcocite	αCu_2S	Portlandite	$Ca(OH)_2$
β Chalcocite	βCu_2S	Pyrrhotite	FeS

74

Mineral	Formula	Mineral	Formula
Chalcopyrite	$CuFeS_2$	Quartz	SiO_2
Copper	Cu	Sepiolite	$2MgO \cdot 3SiO_2 \cdot 2H_2O$
Copper aluminate	$CuAlO_2$	Serpentine	$3MgO \cdot 2SiO_2 \cdot 2H_2O$
Corundum	αAl_2O_3	Siderite	$FeCO_3$
Covellite	CuS	Smithsonite	$ZnCO_3$
Cristroballite	SiO_2	Sodium disilicate	$\beta Na_2Si_2O_5$
Cuprite	Cu_2O	Sodium metasilicate	Na_2SiO_3
Delafossite	$Cu_2O \cdot Fe_2O_3$	Sphalerite	βZnS
Dolomite	$CaCO_3 \cdot MgCO_3$	Syngenite	$K_2SO_4 \cdot CaSO_4 \cdot H_2O$
Favalite	$2FeO \cdot SiO_2$	Talc	$3MgO \cdot 4SiO_2 \cdot H_2O$
Ferrous bicarbonate	$Fe(HCO_3)_2$	Tenorite	CuO
Ferrous sulfate monohydrate	$FeSO_4 \cdot H_2O$	Teschemacherite	NH_4HCO_3
Ferrous sulfate quadrahydrate	$FeSO_4 \cdot 4H_2O$	Thenardite	Na_2SO_4V
δ Ferric Oxide	δFe_2O_3	Thermonatrite	$Na_2CO_3 \cdot H_2O$
β Ferric oxide monohydrate	$\beta Fe_2O_3 \cdot H_2O$	Troilite	FeS
Fluorite	CaF_2	Trona	$3Na_2O \cdot 4CO_2 \cdot 5H_2O$
Foshagite	$5CaO \cdot 3SiO_2 \cdot 3H_2O$	Vermiculite	$(Mg,Fe)_3(Al,Si)_4O_{10}(OH)_2 \cdot 4H_2O$
Forsterite	$2MgO \cdot SiO_2$	Vivianite	$Fe_3(PO_4)_2 \cdot 8H_2O$
Gaylussite	$CaCO_3 \cdot Na_2CO_3 \cdot 5H_2O$	Wedellite	$CaC_2O_4 \cdot 2H_2O$
Gehlenite	$3CaO \cdot Al_2O_3 \cdot 2SiO_2$	Wikeite	$Ca_{10}O[(Si,P,S)O_4]_6$
Gibbsite	$Al_2O_3 \cdot 3H_2O$	Willemite	Zn_2SiO_4
Glauconite	$K_2(MgFe)_2Al_6(Si_4O_{10})_3(OH)_{12}$	Witherite	$BaCO_3$
Goethite	$\alpha Fe_2O_3 \cdot H_2O$	Whitlockite	$\beta Ca_3P_2O_8$
Gypsum	$CaSO_4 \cdot 2H_2O$	Wollastonite	$\beta CaSiO_3$
Gyrolite	$2CaO \cdot 3SiO_2 \cdot H_2O$	Wustite	FeO
Halite	$NaCl$	Xonotlite	$5CaO \cdot 5SiO_2 \cdot H_2O$
Hemihydrate	$CaSO_4 \cdot \frac{1}{2}H_2O$	Zinc A	$ZnO \cdot xH_2O$
Hematite	Fe_2O_3	B	$ZnO \cdot yH_2O$
Hydromagnesite	$4MgO \cdot 3CO_2 \cdot 4H_2O$	E	$ZnO \cdot zH_2O$
Hydrotalcite	$MgCO_3 \cdot 5Mg(OH)_2 \cdot$	Zincite	ZnO
	$2Al(OH)_3 \cdot 4H_2O$	Zincosite	$ZnSO_4$

SOURCE: *Manual on Water*, 3rd ed., ASTM, STP-442, 1969.

Figure 4.5   Friable brown deposit patches on a brass condenser tube. The deposit contained more than 30% manganese.

field by the strong odor of hydrogen sulfide, which is produced after exposure to hydrochloric acid.

Many metal sulfides produce poorly adherent corrosion product layers. This leads to rapid spalling during thermal cycling or turbulent flow. In particular, nonadherent and easily spalled sulfides form on steel and cast irons.

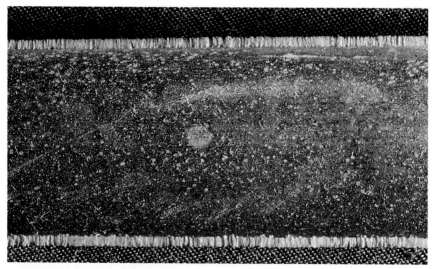

Figure 4.6   A thin, glassy layer of predominantly manganese oxide on the internal surface of a brass condenser tube. The many white spots are pits at fractures in the manganese layer.

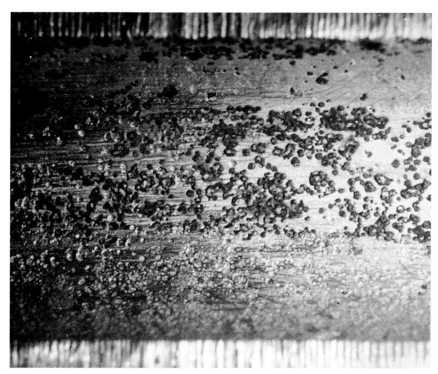

Figure 4.7 Small, manganese-rich nodules on a 90:10 cupronickel condenser tube. Note the small pits beneath each nodule. (*Courtesy of National Association of Corrosion Engineers; Andy Howell Public Service of Colorado, Corrosion '89 Paper No. 197 by H. M. Herro.*)

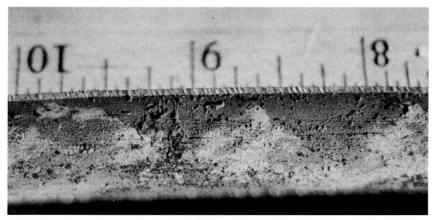

Figure 4.8 Pitting on internal surface of a brass condenser tube revealed by removal of manganese-rich deposit.

Figure 4.9   Tubercles at breaks in a galvanized layer on steel.

Grease and oils

Petroleum greases and oils can be excellent corrosion inhibitors on a variety of alloys. The hydrophobic layer produced by oil or grease can prevent water from contacting surfaces and can, therefore, almost eliminate corrosion. Unfortunately, the addition of oil and grease cannot be recommended as a corrosion-reduction measure in cooling water systems for three basic reasons.

First, greases and oils serve as binders that accumulate and hold large amounts of potentially corrosive substances. Second, greases and

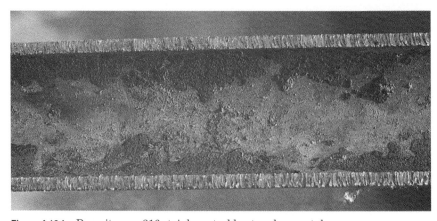

Figure 4.10A   Deposits on a 316 stainless steel heat exchanger tube.

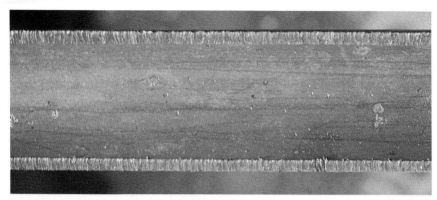

Figure 4.10*B*   As in Fig. 4.10*A*, with deposits removed to show pitting.

Figure 4.11   General wastage on a 304 stainless steel heat exchanger tube.

Figure 4.12   As in Fig. 4.11. Note how abruptly corrosion terminates at deposit edge.

Figure 4.13 Rust patches surround small pits on a stainless steel shaft. The rust is formed by oxidation of ejected ferrous ion from the tiny pits. Pits were initiated by chloride and sulfate ions concentrated by evaporation.

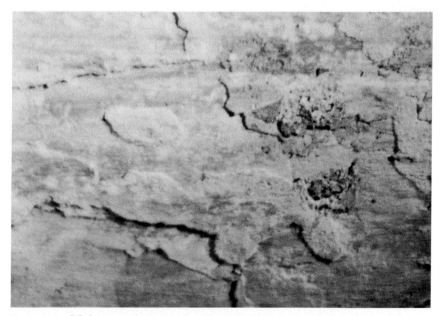

Figure 4.14 Silt layer on the internal surface of a 90:10 cupronickel condenser tube.

Figure 4.15 As in Fig. 4.14 but with silt removed to reveal bluish-white copper carbonate corrosion products.

Figure 4.16 Thick calcium carbonate deposits on condenser tube and copper transfer pipe. Note the stratification.

Figure 4.17   Corroded copper tubes from a heating coil. Cooling water contacted the visible surface. The mounds consist of copper carbonate (see Fig. 4.18).

oils can break down by a variety of processes to produce organic acids that accelerate corrosion. Finally, grease and oil may harbor aggressive microorganisms (see Chap. 6). Attacked surfaces resemble those associated with tuberculation, acid wastage, and microbiologically influenced corrosion.

Elimination

Deposit-related corrosion may be lessened in the following five basic ways.

Deposit removal

Regular cleaning involving mechanical methods is commonly employed. Water blasting, air rumbling, and chemical cleaning often are used as regular maintenance procedures. Cleaning balls made of various materials have been used in condensers and heat exchangers to remove waterside deposits with varying success.

Design changes

Deposition caused by settling of suspended particulate may be reduced by increasing flow. Dead legs, stagnant areas, and other low-flow

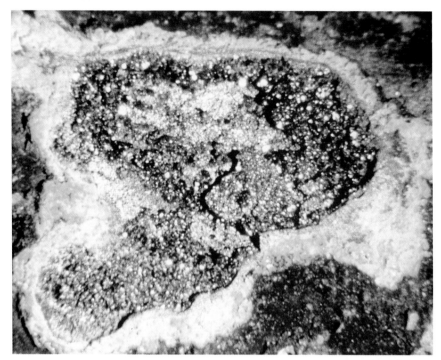

Figure 4.18 As in Fig. 4.17 but with carbonate mound removed to reveal sparkling lavender cuprous oxide crystals.

regions should be eliminated or redesigned to permit flow greater than several feet per second (1 m/s). Flow may be redirected to permit settling of suspended matter in less critical areas of the system.

Water treatment

Removing suspended solids, decreasing cycles of concentration, and clarification all may be beneficial in reducing deposits. Biodispersants and biocides should be used in biofouled systems. Simple pH adjustment may lessen precipitation of certain chemical species. The judicious use of chemical corrosion inhibitors has reduced virtually all forms of aqueous corrosion, including underdeposit corrosion. Of course, the cleaner the metal surface, the more effective most chemical inhibition will be. Process leaks must be identified and eliminated.

Cathodic protection

Applied current devices as well as sacrificial anodes have frequently been used to decrease corrosion associated with deposition. The effec-

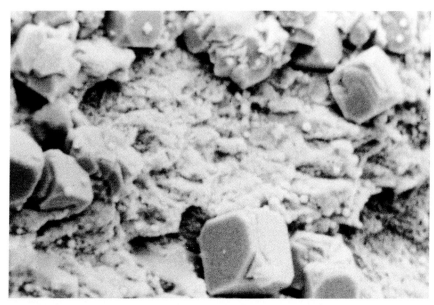

Figure 4.19   Scanning electron microscope picture of cuprous oxide crystals as shown in Fig. 4.18. Note the partial octahedral symmetry.

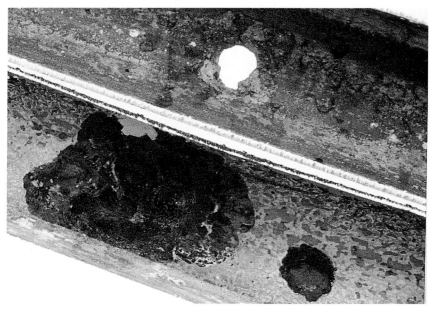

Figure 4.20   Sulfide deposits (dark patches) on longitudinally split brass heat exchanger tube. Note the perforation where wastage penetrated the tube wall. Sulfide was spalled after perforation by escaping fluids.

tiveness and economics of these techniques are directly related to surface cleanliness. In some systems, however, cathodic protection is clearly the most economical technique to lessen attack.

Materials substitution

Sometimes there is no practical, economical way to reduce deposition, protect surfaces cathodically, change design and operation, or treat existing systems chemically. A material change is required. The most economical solution usually is to coat existing structures with water-impermeable, sacrificial, or corrosion-resistant materials.

When surfaces are intermittently wetted, painting sometimes sufficiently reduces corrosion, but it is less effective when constant immersion and deposition occur. Phenolic resins, epoxies, and other organic coatings are quite effective in reducing attack when thermal expansion and contraction are slight and when surface contours are smooth. In more severe service and when temperatures are high, metal spraying and plating may be used.

Substituting one alloy for another may be the only viable solution to a specific corrosion problem. However, caution should be exercised; this is especially true in a cooling water environment containing deposits. Concentration cell corrosion is insidious. Corrosion-resistant materials in oxidizing environments such as stainless steels can be severely pitted when surfaces are shielded by deposits. Each deposit is unique, and nature can be perverse. Thus, replacement materials ideally should be tested in the specific service environment before substitution is accepted.

Cautions

Underdeposit corrosion is not so much a single corrosion mechanism as it is a generic description of wastage beneath deposits. Attack may appear much the same beneath silt, precipitates, metal oxides, and debris. Differential oxygen concentration cell corrosion may appear much the same beneath all kinds of deposits. However, when deposits tend to directly interact with metal surfaces, attack is easier to recognize.

Even when interaction is direct, forms of attack may be difficult to differentiate from other corrosion mechanisms. For example, corrosion beneath biological material can exhibit strikingly unique or vague attack morphologies according to the degree of microbiological involvement (see Chap. 6).

Related Problems

See also Chap. 2, "Crevice Corrosion"; Chap. 3, "Tuberculation"; Chap. 5, "Oxygen Corrosion"; Chap. 6, "Biologically Influenced Corrosion"; and Chap. 13, "Dealloying."

CASE HISTORY 4.1

Industry:	Petroleum refining
Specimen Location:	Hydrocarbon cooler
Specimen Orientation:	Horizontal
Environment:	Shell side: Hydrocarbon 134°F (57°C) inlet, 120°F (49°C) outlet, 16.9 MPa pressure
	Tube side: Cooling water 81°F (27°C) inlet, 109°F (43°C) outlet, 0.4 MPa pressure
	Treatment: Zinc-chromate, 15–20 ppm, dispersant, 0.1–0.2 ppm free Cl_2
	Tower water: Calcium as $CaCO_3$ 450 ppm, Mg as CO_3 100 ppm, chloride 900 ppm, sulfate 150 ppm, total iron 1–2 ppm, M alkalinity as $CaCO_3$ 15, pH 6.5–7.5, conductivity 3500 µmhos, turbidity <20 NTU
Time in Service:	1 year
Sample Specifications:	¾ in. (1.9 cm) outer diameter, 5% chromium, 0.5% molybdenum steel

Severe pitting on the waterside surfaces of a large heat exchanger necessitated replacement of the exchanger tube bundle after only 2 years of service. The exchanger was retubed with a higher chromium alloy in the mistaken belief this metallurgy would be more corrosion resistant. After one more year of service, the new tubes began to fail.

Internal surfaces exhibited many rounded, mutually intersecting pits partially buried beneath silt, iron oxide, and sand deposits. Orange and brown corrosion products and deposits overlaid all. Sulfides were present in the deposits and corrosion products. The material was easily removed when acid was applied (Figs. 4.21 and 4.22).

Corrosion products contained sulfur up to 5% by weight. Aggressive sulfur- and chlorine-containing species concentrated beneath iron oxide, silt, and sand deposits. Localized areas of attack resulted.

Figure 4.21 Internal surface (waterside) of alloy steel heat exchanger. Corrosion products and deposits partially cover pits. (Magnification: 7.5×.)

Figure 4.22 As in Fig. 4.21 but with deposits and corrosion products removed to reveal numerous depressions. (Magnification: 7.5×.)

CASE HISTORY 4.2

Industry:	Fossil utility
Specimen Location:	Main condenser near inlet water box
Specimen Orientation:	Horizontal
Environment:	Tube side: Brackish, estuarine water (polluted), 2–3 ppm tolyltriazole residual, ferrous sulfate ~1 ppm as iron for 2 months for 1 hr/day, dispersant 5–8 ppm
	Shell side: Condensing steam
Time in Service:	2 years
Sample Specifications:	$\frac{7}{8}$ in. (2.2 cm) outer diameter, 90:10 cupronickel tube

The condenser was plagued by rapid attack on waterside surfaces. The entire internal surface was fouled with silt and other deposits, beneath which a cuprous oxide layer was present (Fig. 4.23). Localized areas of metal loss were present beneath mounds of corrosion product. Some of these localized areas were deep enough to threaten tube integrity.

Ferrous sulfate was added to the condenser in hopes of retarding attack. A thick, tan deposit layer rapidly formed on tubes near inlets (Fig. 4.24). Corrosion continued unabated. Underdeposit corrosion caused localized areas of metal loss (Fig. 4.25). Corrosion product mounds contained up to 10% chloride.

More frequent manual cleaning of tubes was suggested as on-line sponge-ball cleaning did not appear effective. Higher concentrations of tolyltriazole were fed with no discernible effect.

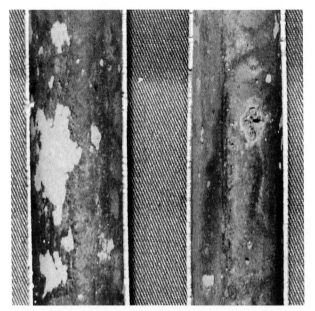

Figure 4.23 Thick, red cuprous oxide layer covering internal surface of 90:10 cupronickel condenser tube. Note the green mounds marking sites of localized wastage.

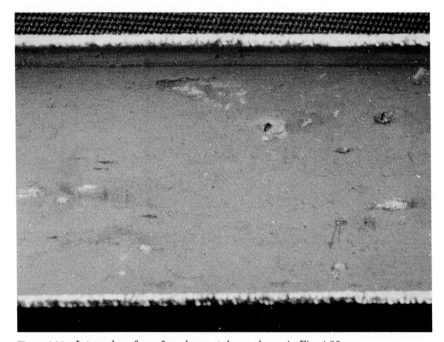

Figure 4.24 Internal surface of condenser tube as shown in Fig. 4.23.

Figure 4.25   As in Figs. 4.23 and 4.24. Corrosion product mound broken open to reveal sparkling cuprous oxide crystals.

CASE HISTORY 4.3

Industry:	Refining
Specimen Location:	Flexicoker heat exchanger
Specimen Orientation:	Horizontal
Environment:	Internal: Temperature up to 190°F (88°C) at outlet, flow 1–5 ft/s (0.3–1.5 m/s), pH 7.5–8.0, hardness 100–400 ppm, conductivity 4000 μmhos max., M alkalinity 100–200 ppm; treatment 8–10 ppm phosphate, 0.5 ppm zinc soluble min., antifoam as needed; high-turbidity makeup water (mud to silt)
	External: 250°F (121°C) max.
Time in Service:	5 years
Sample Specifications:	¾ in. (1.9 cm) outer diameter, carbon steel; 0.08 in. (0.20 cm) wall thickness

The heat exchanger had a history of fouling with silt and precipitates. Water pH was decreased to reduce fouling, with some success.

Tubes were removed for routine inspection. Internal surfaces had rough contours due to mutually intersecting areas of metal loss (Fig. 4.26). Wall thickness varied from 0.080 in. (0.20 cm) to as little as 0.032 in. (0.081 cm).

Some of the observed wastage was caused by past acid cleanings. However, much of the attack was caused by long-term underdeposit corrosion.

Figure 4.26 Internal surface of steel heat exchanger tube after removal of deposits. Note the mutually intersecting areas of metal loss.

CASE HISTORY 4.4

Industry:	Primary metals
Specimen Location:	Corrosion coupon from mill coolant tank
Specimen Orientation:	Vertical
Environment:	Mill coolant temperature ~160°F (71°C), turbulent flow
Time in Service:	6 months
Sample Specifications:	3 in. by 5 in. by ¹⁄₁₆ in. (7.6 cm by 12.7 cm by 0.16 cm), cold-rolled 1010 steel

Carbon steel contacting mill coolant had suffered general corrosion. Stainless steel components were unaffected. Although many factors contributed to wastage in these systems, deposits played an important role (Fig. 4.27A and B). Corrosion exactly mirrored deposition patterns.

Deposits contained organic acids formed by oxidation of rolling oils. Up to 40% by weight of the lumps shown in Fig. 4.27A and B was iron oxides, hydroxides, and organic-acid iron salts. Acidic species concentrated in the deposits.

After increasing coolant corrosion-inhibitor concentration, coupon corrosion rates decreased by almost 70%.

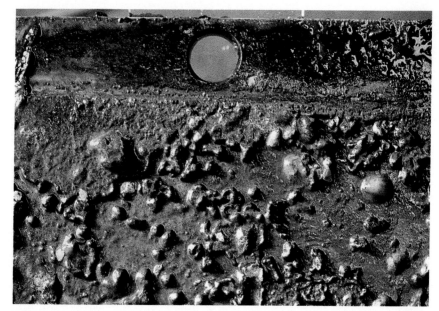

Figure 4.27*A*   Mild steel coupon removed from a rolling mill cooling tank. Note the thick greasy deposits.

Figure 4.27*B*   As in Fig. 4.27*A* but with deposits removed to show corrosion beneath.

CASE HISTORY 4.5

Industry:	Refinery
Specimen Location:	Depropanizer overhead condenser, top row of tubes, middle of tube bundle
Specimen Orientation:	Horizontal
Environment:	Internal: Untreated well water, 70–80°F (21–27°C), pressure 10–20 psi (0.2–0.4 MPa), pH 7.0–7.5, chloride 120 ppm, bicarbonate alkalinity 224 ppm, sulfate 83 ppm, conductivity 921 μmhos, turbidity 2 NTU
	External: Depropanizer overhead gas, 90°F (32°C); trace water, mercaptans, ammonia, H_2S; chlorides 130–140 ppm, pressure 200–300 psi (1.4–2.1 MPa)
Time in Service:	17 years
Sample Specifications:	¾ in. (1.9 cm) outer diameter duplex tubes, admiralty brass internally, aluminum externally

A duplex heat exchanger tube containing a single small perforation was examined. Perforation occurred due to internal surface wastage beneath a deposit layer containing large concentrations of sulfur compounds (Fig. 4.28).

Corrosion proceeded generally on internal surfaces during a long period. Eventually, the inner brass tube was penetrated, exposing the aluminum tube to internal cooling waters. The aluminum corroded rapidly at the breach, since coupling of brass and aluminum produced a galvanic cell (see Chap. 16, "Galvanic Corrosion").

The relatively long service life of the duplex tubes suggested that similar tubes be used for retubing. Because of supplier delays in securing duplex tubing, however, alternative materials, including admiralty brass, were considered.

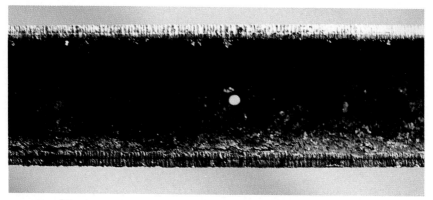

Figure 4.28   Perforation on a duplex (brass internally, aluminum externally) tube caused by sulfur compounds concentrated on internal surfaces.

CASE HISTORY 4.6

Industry:	University building
Specimen Location:	Air-conditioning chilled-water coil
Specimen Orientation:	Horizontal
Environment:	Internal: 55–43°F (13–6°C), pressure 30 psi (0.2 MPa) max., pH 8.5–9.5, M alkalinity 70 ppm, calcium 60 ppm, silica 7 ppm, Fe trace; treatment molybdate 9–20 ppm, *previously* zinc-chromate
	External: Ambient air
Time in Service:	12–18 years
Sample Specifications:	½ in. (1.3 cm) outer diameter copper tubes, embedded in a matrix of thin aluminum heat exchanger fins

It was reported that makeup rates due to tube leaks, bad pump packing, and other losses was up to 70,000 gal/hr in this 2 million gallon system, although the design was for 10,000 gal/hr makeup.

Internal surfaces of all tubes were severely attacked (Fig. 4.29). A brown deposit layer consisting of magnetite, iron oxide hydroxide, and silica covered all surfaces. Deposition was thicker and more tenacious along the bottom of tubes. These deposits had a distinct greenish-blue cast caused by copper corrosion products beneath the deposit. Underlying corrosion products were ruby-red cuprous oxide crystals (Fig. 4.29). Areas not covered with deposits suffered only superficial attack, but below deposits wastage was severe.

Wastage was caused by classic long-term underdeposit corrosion. The combined effects of oxygen concentration cells, low flow, and contamination of system water with high chloride- and sulfate-concentration makeup waters caused corrosion.

Figure 4.29   Corroded internal surface of a copper chiller tube. The light-colored deposits are primarily iron oxides and silicates; reddish material is cuprous oxide.

CASE HISTORY 4.7

Industry:	Natural gas
Specimen Location:	Main gas compressor third-stage cooler
Specimen Orientation:	Horizontal
Environment:	Internal: Water contaminated with sulfur compounds, 70–90°F (21–32°C)
	External: 270–100°F (132–38°C), pressure 520 psig (3.6 MPa), 7.6% N_2, 2–3% O_2, 2–3% CO_2, remainder other hydrocarbons
Time in Service:	9 months
Sample Specifications:	1 in. (2.5 cm) outer diameter, admiralty brass tubes

Tubes were almost plugged with deposits (Fig. 4.30A and B). The deposits contained almost 40% sulfur by weight. Hydrocarbon in-leakage had occurred through small cracks that originated on internal surfaces beneath deposits.

Cracking was caused by stress-corrosion cracking (see Chap. 9, "Stress-Corrosion Cracking") involving hydrogen sulfide and/or moist sulfur dioxide. The sulfur entered the cooling water stream through process leaks, which were repaired.

Figure 4.30*A*   Brass cooler tube almost plugged by deposits.

Figure 4.30*B*   As in Fig. 4.30A. Note the light-colored deposits intermixed with bluish-green corrosion products.

Oxygen Corrosion

General Description

Iron and steel

Water of neutral pH containing no dissolved oxygen does not significantly corrode iron or steel near room temperature. The corrosion rate of iron in deaerated water at room temperature is less than 0.2 mil/y (0.005 mm/y). In air-saturated water, corrosion rates can be 100 times greater. The corrosion rate of iron and most nonstainless alloys is nearly proportional to the concentration of dissolved oxygen (at least at oxygen concentrations up to several parts per million) (Fig. 5.1). Corrosion ceases when all dissolved oxygen is consumed. In fact, deaeration has been attempted by passing water over sacrificial steel parts having large surface-to-volume ratios. The water is stripped of oxygen by corrosion with the steel and thus is less corrosive to steel downstream.

Corrosion rates in an airtight system containing air-saturated water can be calculated (assuming only Fe_2O_3 is formed as a corrosion product). Average metal loss would be about 0.0002 in. (0.0006 cm) deep in a spherical tank containing 100,000 ft^3 (2830 m^3) of air-saturated water at room temperature if all the dissolved oxygen was consumed. It should be obvious that appreciable oxygen corrosion requires very large amounts of water contacting metal surfaces for prolonged periods. Such conditions are met in a variety of cooling water applications.

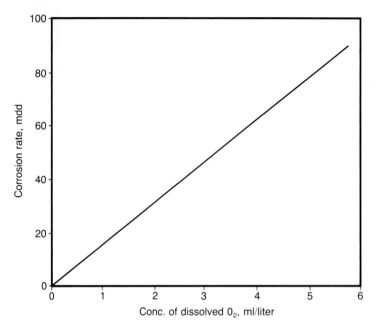

Figure 5.1 Effect of oxygen concentration on corrosion of mild steel in slowly moving water containing 165 ppm $CaCl_2$; 48-hour test, 25°C. [*Courtesy of H. H. Uhlig, D. N. Triadis, and M. Stern, "Effect of Oxygen, Chlorides, and Calcium Ion on Corrosion Inhibition of Iron by Polyphosphates," J. Electrochem. Soc. 102, p. 60 (1955). Reprinted with permission by The Electrochemical Society, Inc.*]

The major cathodic and anodic reactions in near-neutral pH water are shown in Reactions 5.1 to 5.3:

Anode: $$Fe \rightarrow Fe^{++} + 2e^- \tag{5.1}$$

Cathode: $$2H^+ + \tfrac{1}{2}O_2 \rightarrow H_2O - 2e^- \tag{5.2}$$

(predominates in aerated, alkaline solutions)

$$H^+ \rightarrow \tfrac{1}{2}H_2 - e^- \tag{5.3}$$

(predominates in deaerated, acid solutions)

The corrosion rate is controlled mainly by cathodic reaction rates. Cathodic Reactions 5.2 and 5.3 are usually much slower than anodic Reaction 5.1. The slower reaction controls the corrosion rate. If water pH is depressed, Reaction 5.3 is favored, speeding attack. If oxygen concentration is high, Reaction 5.2 is aided, also increasing wastage by a process called *depolarization*. Depolarization is simply hydrogen-ion removal from solution near the cathode.

Reaction 5.2 is controlled primarily by the diffusion of oxygen to the corroding surface. Any process that slows oxygen diffusion slows Reaction 5.2, ultimately reducing corrosion rates. As corrosion products accumulate on the corroding surface, oxygen diffusion and Reaction 5.2 are slowed, resulting in a decreased corrosion rate. It is precisely for this reason that corrosion coupons should be exposed in oxygenated water long enough to achieve reproducible corrosion-rate data.

When a clean steel coupon is placed in oxygenated water, a rust layer will form quickly. Corrosion rates are initially high and decrease rapidly while the rust layer is forming. Once the oxide forms, rusting slows and the accumulated oxide retards diffusion. Thus, Reaction 5.2 slows. Eventually, nearly steady-state corrosion is achieved (Fig. 5.2). Hence, a minimum exposure period, empirically determined by the following equation, must be satisfied to obtain consistent corrosion-rate data for coupons exposed in cooling water systems (Figs. 5.2 and 5.3):

$$\text{Minimum exposure time*} = \frac{85}{\text{corrosion rate}}$$

The corrosion-product layer that forms due to oxygen corrosion is discussed in Chap. 3, "Tuberculation." However, the initial corrosion product is ferrous hydroxide [Fe $(OH)_2$] (Reaction 5.4):

$$Fe + H_2O + \tfrac{1}{2}O_2 \rightarrow Fe\,(OH)_2 \qquad (5.4)$$

---

* Minimum exposure time is calculated as number of days; corrosion rate is in milliinches per year.

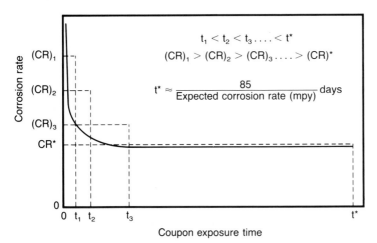

Figure 5.2 Schematic of carbon steel corrosion rate versus exposure time in a typical oxygenated cooling water. Note how the average corrosion rate decreases with time and converges to CR* at t^* (the minimum exposure time to get reproducible results).

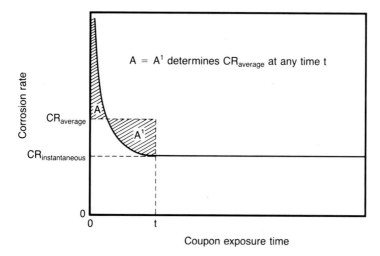

Figure 5.3 As in Fig. 5.2. Note that the average corrosion rate (CR average) is always larger than the instantaneous corrosion rate (CR instantaneous) and that this difference is largest at small exposure times.

The ferrous hydroxide is rapidly oxidized to ferric hydroxide in oxygenated waters (Reaction 5.5):

$$\text{Fe (OH)}_2 + \tfrac{1}{2}\text{H}_2\text{O} + \tfrac{1}{4}\text{O}_2 \rightarrow \text{Fe (OH)}_3 \tag{5.5}$$

Ferric and ferrous hydroxide usually contain water of hydration. A layer of hydrated magnetite sometimes forms between the ferric and ferrous hydroxides (see Fig. 3.2). The threefold layer of corrosion products is usually called *rust*. The bulk of rust is the ferric hydroxide layer. (In tubercles, however, ferrous hydroxide often is the major component; see Chap. 3.)

Oxygen corrosion of steel doubles for every 35–55°F (20–30°C) rise in temperature, beginning near room temperature. Corrosion is nearly proportional to temperature up to about 180°F (80°C) in systems open to the air. Although reaction rates increase with temperature, dissolved oxygen is driven from solution as temperatures increase. As temperatures approach boiling, corrosion rates fall to very low values, since dissolved-oxygen concentration also decreases as water temperature rises (Fig. 5.4).

When water pH is between about 4 and 10 near room temperature, iron corrosion rates are nearly constant (Fig. 5.5). Below a pH of 4, protective corrosion products are dissolved. A bare iron surface contacts water, and acid can react directly with steel. Hydrogen evolution (Reaction 5.3) becomes pronounced below a pH of 4. In conjunction with oxygen depolarization, the corrosion rate increases sharply (Fig. 5.5).

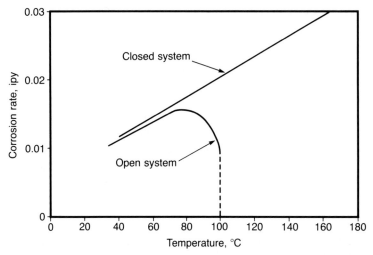

Figure 5.4   Effect of temperature on corrosion of iron in water containing dissolved oxygen. (*Courtesy of McGraw-Hill Inc., Corrosion, Causes and Prevention, by F. Speller, p. 168, McGraw-Hill, 1951*)

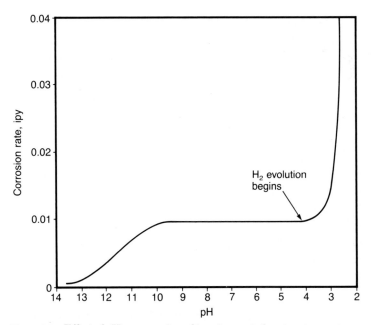

Figure 5.5   Effect of pH on corrosion of iron in aerated water at room temperature. (*Reprinted from* Industrial & Engineering Chemistry *16, 665, 1924, W. Whitman, R. Russell, and V. Altieri, published by the American Chemical Society.*)

Above a pH of about 10 near room temperature, surface alkalinity increases and corrosion rates fall sharply. When sodium-hydroxide concentration rises to several percent, the corrosion rate drops to almost zero. Even at very high sodium-hydroxide concentrations, the corrosion rate increases only slightly.

Differences in alloy carbon concentration, heat treatment, and mechanical forming usually produce only small differences in corrosion rate in a pH range of 4–10. It is less certain how corrosion rates vary at high and low pH due to these factors. Cast irons containing graphite particles may experience a unique form of attack called *graphitic corrosion* (see Chap. 17, "Graphitic Corrosion").

Copper alloys

Copper is not ordinarily corroded in water unless dissolved oxygen is present. In nearly pure aerated water, a thin, protective layer of cuprous oxide and cupric hydroxide forms. Oxygen must diffuse through the film for corrosion to occur.

Oxygen has a dual role in corrosion of copper. It stimulates attack by an interaction at the cathode, and it retards corrosion by producing a protective corrosion-product layer. High velocity can disturb the corrosion-product layer, increasing attack.

Dissolved carbon dioxide, acids, chlorides, sulfates, ammonia, sulfur dioxide, and sulfides are corrosive to copper alloys. Alloying with zinc reduces attack by waters containing hydrogen sulfide; higher-zinc brasses are the most resistant. Dissolved oxygen must be present for sulfuric acid corrosion. In sea water, corrosion rates of brasses containing at least 68% copper are low. Corrosion rates of copper in slow-moving sea water can be lower than 0.002 mil/y (0.05 mm/y). However, corrosion rates in sodium chloride solutions increase strongly with temperature, almost doubling for each 68°F (20°C) rise in temperature up to about 176°F (80°C). As water temperature increases, the tendency for localized attack increases. Increased concentration of hydrogen sulfide has been correlated with increased rates of pitting.

In most respects, copper-nickel and copper-tin alloys behave similarly to copper-zinc alloys. The presence of acids, hydrogen sulfide, ammonia, and carbon dioxide degrades corrosion resistance.

Aluminum

Aluminum alloys are essentially unaffected by dissolved oxygen in pure water up to 350°F (180°C). Although much of aluminum's corrosion resistance is due to the presence of an adherent oxide film, oxygen is not necessary to form the layer. Direct reaction with water can produce the protective oxide. Aluminum is often used to contain distilled water for the above reasons.

Attack is usually the direct result of high or low pH conditions in the presence of aggressive anions such as chloride.

Stainless steels

Stainless steels contain 11% or more chromium. Table 5.1 lists common commercial grades and compositions of stainless steels. It is chromium that imparts the "stainless" character to steel. Oxygen combines with chromium and iron to form a highly adherent and protective oxide film. If the film is ruptured in certain oxidizing environments, it rapidly heals with no substantial corrosion. This film does not readily form until at least 11% chromium is dissolved in the alloy. Below 11% chromium, corrosion resistance to oxygenated water is almost the same as in unalloyed iron.

Corrosion resistance of stainless steel is reduced in deaerated solutions. This behavior is opposite to the behavior of iron, low-alloy steel, and most nonferrous metals in oxygenated waters. Stainless steels exhibit very low corrosion rates in oxidizing media until the solution oxidizing power becomes great enough to breach the protective oxide locally. The solution pH alone does not control attack (see Chap. 4, "Underdeposit Corrosion"). The presence of chloride and other strong "depassivating" chemicals deteriorates corrosion resistance.

Zinc

The principal metal used in galvanizing coatings is zinc. In cooling water systems, use of galvanized panels and steel pipe is common.

Zinc corrodes at a rate somewhere between that of iron and copper in most natural waters. Voluminous oxides and hydroxides form when zinc is immersed in oxygenated water. Attack will occur in the absence of oxygen, much as on aluminum, but corrosion rates are as low as 1 mil/y (0.025 mm/y) at room temperature when no oxygen is present; corrosion rates are three times greater at 149°F (65°C). When water is saturated with oxygen, corrosion rates are about 9 mil/y (0.23 mm/y) at room temperature; rates increase to 14 mil/y (0.36 mm/y) at 104°F (40°C). Attack is usually general if oxygen supply is adequate. Localized attack is favored when oxygen concentration falls below saturation levels.

Zinc is amphoteric; it is corroded at high and low pH. At 86°F (30°C), corrosion rates are as high as 190 mil/y (4.8 mm/y) at a pH of 3 and are 70 mil/y (1.8 mm/y) at a pH of 13.5. The corrosion rate is as little as 2 mil/y (0.05 mm/y) at weakly alkaline pHs.

TABLE 5.1 Summary of Stainless Types and Chemical Compositions

Uniloy type	C (max.)	Mn (max.)	P (max.)	S (max.)	Si (max.)	Cr	Ni	Mo (max.)	Other elements
					Austenitic Types				
201	0.15	5.50–7.50	0.060	0.030	1.00	16.00–18.00	3.50–5.50	—	N 0.25 max.
202	0.15	7.50–10.00	0.060	0.030	1.00	17.00–19.00	4.00–6.00	—	N 0.25
301	0.15	2.00	0.045	0.030	1.00	16.00–18.00	6.00–8.00	—	—
302	0.15	2.00	0.045	0.030	1.00	17.00–19.00	8.00–10.00	—	—
303	0.15	2.00	0.20	0.15 min.	1.00	17.00–19.00	8.00–10.00	0.60	—
303-Se	0.15	2.00	0.20	0.060	1.00	17.00–19.00	8.00–10.00	—	Se 0.15 min.
303MA	0.15	2.00	0.040	0.11–0.16	1.00	17.00–19.00	8.00–10.00	0.40–0.60	Al 0.60–1.00
304	0.08	2.00	0.045	0.030	1.00	18.00–20.00	8.00–10.00	—	—
304L	0.030	2.00	0.045	0.030	1.00	18.00–20.00	8.00–10.00	—	—
305	0.12	2.00	0.045	0.030	1.00	17.00–19.00	10.00–13.00	—	—
308	0.08	2.00	0.045	0.030	1.00	19.00–21.00	10.00–12.00	—	—
309	0.20	2.00	0.045	0.030	1.00	22.00–24.00	12.00–15.00	—	—
309S	0.08	2.00	0.045	0.030	1.00	22.00–24.00	12.00–15.00	—	—
310	0.25	2.00	0.045	0.030	1.50	24.00–26.00	19.00–22.00	—	—
310S	0.08	2.00	0.045	0.030	1.50	24.00–26.00	19.00–22.00	—	—
314	0.25	2.00	0.045	0.030	1.50–3.00	23.00–26.00	19.00–22.00	—	—
316	0.08	2.00	0.045	0.030	1.00	16.00–18.00	10.00–14.00	2.00–3.00	—
316L	0.030	2.00	0.045	0.030	1.00	16.00–18.00	10.00–14.00	2.00–3.00	—
316F	0.08	2.00	0.10 min.	0.10 min.	1.00	17.00–19.00	10.00–14.00	1.75–3.00	—
317	0.08	2.00	0.045	0.030	1.00	18.00–20.00	11.00–15.00	3.00–4.00	—
317L	0.030	2.00	0.045	0.030	1.00	18.00–20.00	11.00–15.00	3.00–4.00	—
318	0.08	2.00	0.045	0.030	1.00	17.00–19.00	13.00–15.00	2.00–3.00	Cb 10XC min.
D319	0.07	2.00	0.045	0.030	1.00	17.50–19.50	11.00–15.00	2.25–3.00	—
321	0.08	2.00	0.045	0.030	1.00	17.00–19.00	9.00–12.00	—	Ti 5XC min.
330	0.20	2.00	0.030	0.030	1.50	14.50–16.50	34.00–37.00	—	—
347	0.08	2.00	0.045	0.030	1.00	17.00–19.00	9.00–13.00	—	Cb-Ta 10XC min.
347-Se	0.08	2.00	0.10 min.	0.030	1.00	17.00–19.00	9.00–12.00	—	Se 0.15 min.
348	0.08	2.00	0.045	0.030	1.00	17.00–19.00	9.00–13.00	—	Cb-Ta 10XC min. Ta 0.10 max. Co 0.20 max.
16–18	0.08	2.00	0.045	0.030	1.00	15.00–17.00	17.00–19.00	—	—

Hardenable Types

Type									
403	0.15	1.00	0.040	0.030	0.50	11.50–13.00	—	—	—
410	0.15	1.00	0.045	0.030	1.00	11.50–13.50	—	—	—
414	0.15	1.00	0.045	0.030	1.00	11.50–13.50	1.25–2.50	—	—
416	0.15	1.25	0.060	0.15 min.	1.00	12.00–14.00	—	0.60	—
420	0.15 min.	1.00	0.040	0.030	1.00	12.00–14.00	—	—	—
420F	0.15 min.	1.00	0.040	0.15 min.	1.00	12.00–14.00	—	0.60	—
420F-Se	0.15 min.	1.00	0.040	0.030	1.00	12.00–14.00	—	0.60	Se 0.15 min.
431	0.20	1.00	0.040	0.030	1.00	15.00–17.00	1.25–2.50	—	—
440A	0.60–0.75	1.00	0.040	0.030	1.00	16.00–18.00	—	0.75	—
440B	0.75–0.95	1.00	0.040	0.030	1.00	16.00–18.00	—	0.75	—
440C	0.95–1.20	1.00	0.040	0.030	1.00	16.00–18.00	—	0.75	—
440F	0.95–1.20	1.00	0.040	0.05 min.	1.00	16.00–18.00	—	0.75	—
440F-Se	0.95–1.20	1.00	0.040 min.	0.030	1.00	16.00–18.00	—	0.75	Se 0.10 min.

Nonhardenable Types

Type									
405	0.08	1.00	0.040	0.030	1.00	11.50–14.50	—	—	Al 0.10–0.30
430	0.12	1.00	0.040	0.030	1.00	14.00–18.00	—	—	—
430F	0.12	1.25	0.060	0.15 min.	1.00	14.00–18.00	—	0.60	—
430F-Se	0.12	1.25	0.060	0.060	1.00	14.00–18.00	—	—	Se 0.15 min.
430 Ti	0.12	1.00	0.040	0.030	1.00	16.00–18.00	—	—	Ti 6XC min.
434	0.12	1.00	0.040	0.030	1.00	14.00–18.00	—	0.75–1.25	—
435	0.12	1.00	0.040	0.030	1.00	14.00–18.00	—	—	Cb+Ta 0.40–0.60
436	0.12	1.00	0.040	0.030	1.00	14.00–18.00	—	0.75–1.25	Cb+Ta 0.40–0.60
442	0.20	1.00	0.040	0.030	1.00	18.00–23.00	—	—	—
443	0.20	1.00	0.040	0.030	1.00	18.00–23.00	—	—	Cu 0.90–1.25
446	0.20	1.50	0.040	0.030	1.00	23.00–27.00	—	—	N 0.25 max.

SOURCE: Courtesy of Bridgeville Stainless Division, Cyclops Industries, Inc.

Locations

Oxygen corrosion only occurs on metal surfaces exposed to oxygenated waters. Many commonly used industrial alloys react with dissolved oxygen in water, forming a variety of oxides and hydroxides. However, alloys most seriously affected are cast irons, galvanized steel, and non-stainless steels. Attack occurs in locations where tuberculation also occurs (see Chap. 3). Often, oxygen corrosion is a precursor to tubercle development.

Carbon steel heat exchangers, cast iron water boxes, screens, pump components, service water system piping, standpipes, fire protection systems, galvanized steel, engine components, and virtually all non-stainless ferrous components are subject to significant corrosion in oxygenated water.

Identification

Corrosion in heavily oxygenated waters tends to produce general wastage on many alloys (Figs. 5.6 and 5.7). Localized attack, including pitting, is sometimes favored when a high dissolved-oxygen concentration falls to low values. Localized corrosion can also occur when protective corrosion-product layers, formed under favorable conditions,

Figure 5.6   Rust layer on carbon steel mill water supply pipe. Note the partially exfoliated region at the bottom of the photo.

are exposed to elevated oxygen concentrations during upsets or idle periods (Fig. 5.8 and Case History 5.1).

Iron and steel

Corrosion products are principally friable brown and orange rust. Darker, denser, more reduced oxides and hydroxides are often present beneath the more friable layers. Oxides formed at low oxygen concentrations and at high temperatures tend to be dark. Oxides lighten upon exposure to air, becoming yellowish-brown after drying. Those oxides containing high concentrations of chloride and sulfate tend to become red when dried in air. Frequently, tubercles develop if surfaces are constantly immersed (see Chap. 3).

Stratified oxide layers form in misted areas (Figs. 5.6 and 5.7). Stratification of corrosion products is common in intermittent wet-dry conditions (Fig. 5.9). Wastage is usually general if oxygen concentrations remain near saturation at room temperature. Surfaces have irregular, undulating contours formed by shallow, intersecting areas of metal loss (Fig. 5.10). As water pH falls, surface contours become rougher.

Pitting can occur when normally protective corrosion-product or deposit layers are locally breached. Localized attack occurs during upsets or when protracted idle periods change water conditions abruptly. Regions adjacent to localized corrosion sites often remain

Figure 5.7 Rust chip as in Fig. 5.6.

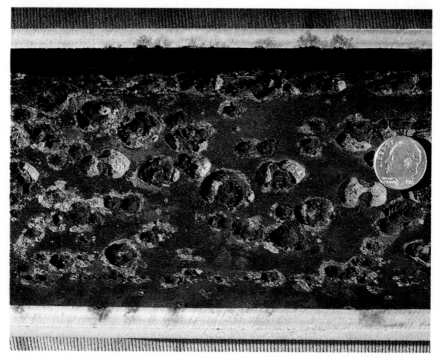

Figure 5.8   Heavily pitted surface of a mild steel cooling water return line. Note the deep localized metal loss surrounded by aureoles of rust. Pitting is common when oxygen concentration varies greatly over time.

smooth and unattacked. Deposits and corrosion products cover smooth metal surfaces, except where attack has occurred. If flow is slow when pitting occurs, a ring of orange and brown precipitated rust usually surrounds each pit (Fig. 5.11). Rust forms from the iron ion that is oxidized upon ejection from the pit to form a ferric hydroxide ridge. If flow is significant, these rings may be absent or elongated into ellipses with major axes paralleling flow direction. Uniform rusting is favored when corrosion is slow. Rapid attack tends to produce a more irregular corrosion-product layer.

Galvanized steel

Zinc is susceptible to attack from oxygen concentration cells. Shielded areas or areas depleted in oxygen concentration tend to corrode, forming voluminous, white, friable corrosion products. Once the zinc layer is breached, the underlying steel becomes susceptible to attack and is severely wasted locally (Figs. 5.12 and 5.13).

Figure 5.9 Wall of steel tank, originally ½ in. (1.3 cm) thick, that was entirely converted to oxide. It was continually exposed to oxygenated water mist at about 180°F (82°C).

Figure 5.10 Generally corroded steel surface.

Figure 5.11   As in Fig. 5.8. Note the rust halos surrounding each pit. Also note how little attack is present between pits.

Figure 5.12   Severely corroded galvanized steel pipe supplying water to a cooling tower fan-bearing system. Where the galvanized metal has been consumed, only a brown rust patch is visible.

Figure 5.13   As in Fig. 5.12 but with pipe in cross section, showing severe wastage where the galvanized layer is absent.

Critical Factors

Differential aeration stimulates localized attack (see Chap. 2, "Crevice Corrosion"). At zero oxygen concentration, oxygen corrosion ceases. If very low oxygen-concentration regions are in contact with regions of appreciable oxygen concentration, however, wastage in the oxygen deficient regions may be severe. It is for this reason that deposits, although not innately aggressive, may stimulate attack by establishing oxygen concentration cells between occluded regions below deposits and adjacent to oxygen-rich regions. Local corrosion rates may be much higher than general corrosion rates due to unfavorable anode-to-cathode area ratios established by differential aeration (see Chap. 4).

Elimination

Oxygen corrosion may be reduced or eliminated by the following:

- Appropriate chemical treatment of water (with corrosion inhibitors, dispersants, and filmers)
- Coating metal surfaces with water-impermeable barriers (such as paint, epoxies, thermal sprays, grease, and oil)

- Substituting more-resistant materials such as stainless steels and copper alloys for less-resistant alloys such as carbon steels
- Deaeration (which may include mechanical, thermal, and/or chemical treatments)
- Minimization of deposits
- Cathodic protection (involving the use of applied current or sacrificial anodes)
- Preventing surfaces from contacting water

Cautions

Oxygen corrosion involves many accelerating factors such as the concentration of aggressive anions beneath deposits, intermittent operation, and variable water chemistry. How each factor contributes to attack is often difficult to assess by visual inspection alone. Chemical analysis of corrosion products and deposits is often beneficial, as is more detailed microscopic examination of corrosion products and wasted regions.

Related Problems

See also Chap. 2, "Crevice Corrosion"; Chap. 3, "Tuberculation"; and Chap. 4, "Underdeposit Corrosion."

CASE HISTORY 5.1

Industry:	Chemical process
Specimen Location:	Main cooling water return header
Specimen Orientation:	Horizontal
Environment:	Water treatment was chrome-zinc for the first 6 years and orthophosphate for the next 5 years. Free residual chlorine 0.1–0.3 ppm; filtered orthophosphate 10–17 ppm, 6–8 cycles; calcium <200 ppm, 200°F (93°C), pressure 50 psi (0.3 MPa).
Time in Service:	11 years
Sample Specifications:	3.5 in. (8.9 cm) outer diameter, mild steel pipe

Steel pipe in a cooling water return header contained numerous deep pits on internal surfaces; one pit penetrated the pipe wall (Fig. 5.8). Pits range in size from pinpoint depressions to ½-in. (1.3-cm) pockets. Some pits are filled with friable, dark oxides and are surrounded with orange oxide rings (Fig. 5.11). Pits are more numerous and deeper on the apparent bottom of the pipe. Surfaces between pits are covered with a tenacious, black oxide layer.

The system was operated cyclically. Steam was supplied to a process reactor to start an exothermic chemical reaction. Once the reaction began, cooling water was supplied to maintain process temperature. Cooling water at 200°F (93°C) entered the return header after exiting the process reactor.

Pitting was caused by intermittent exposure of pipe surfaces to highly oxygenated water. High-concentration episodes were followed by periods of relatively low oxygen concentration during which a protective, black magnetite layer was formed. Weak points in the magnetite layer were attacked during the high-oxygen-concentration excursions, causing intense localized corrosion.

CASE HISTORY 5.2

Industry:	Refinery
Specimen Location:	Two sections of overhead condenser tubing
Specimen Orientation:	Horizontal
Environment:	Internal: Entering water temperature 85°F (29°C), exiting water temperature 115°F (46°C), pressure ~120 psig (0.82 MPa), pH 7.5–7.6, gaseous chlorination 0.3–0.6 ppm free with spike of 3 ppm weekly; 14–16 ppm orthophosphate; 3.0–3.5 ppm zinc; dispersant.
	External: Cracked gasoline, 130 psi (0.9 MPa)
Time in Service:	1 and 12 years
Sample Specifications:	1 in. (2.5 cm) outer diameter, 0.110 in. (0.280 cm) minimum wall thickness, SA214 steel, U-tubes

Two sections of steel condenser tubing experienced considerable metal loss from internal surfaces. An old section contained a perforation; the newer section had not failed. A stratified oxide and deposit layer overlaid all internal surfaces (Fig. 5.14). Corrosion was severe along a longitudinal weld seam in the older section (Fig. 5.15). Differential oxygen concentration cells operated beneath the heavy accumulation of corrosion products and deposits. The older tube perforated along a weld seam.

X-ray analysis of corrosion products and deposits removed from internal surfaces showed 68% iron, 12% phosphorus, 8% silicon, 3% sulfur, and 2% each of zinc, sodium, chromium, and calcium; other materials made up the remainder of deposits and corrosion products.

Figure 5.14   Stratified corrosion products and deposits on the internal surface of a tube. (Magnification: 7.5×.)

Figure 5.15   Rough internal surface contour of the tube shown in Fig. 5.14. Note the preferential attack on the weld seam. (Magnification: 7.5×.)

CASE HISTORY 5.3

Industry:	Chemical process
Specimen Location:	Heat exchanger tubing from a cooling tower
Specimen Orientation:	Vertical
Environment:	~60–80°F (16–27°C), pH 7.5–8.2, chlorination >0.1 free residual
Time in Service:	Unknown
Sample Specifications:	¾ in. (1.9 cm) outer diameter steel tubing, phenolic epoxy resin lined

A 1-ft. (30-cm)-long section of steel heat exchanger tubing containing no failure was received. The section was submitted for evaluation of the internal surface, which was lined with a phenolic epoxy resin.

Close visual examination of internal surfaces using a low-power stereomicroscope revealed a coating of reddish iron oxides on the internal surface. A population of small, knoblike mounds of corrosion product resembling tubercles was present on the surface (Fig. 5.16).

Reexamination of the surface after a brief acid cleaning showed that about 90% was still covered by resin, although the resin had been blistered and ruptured at small tubercular sites (Fig. 5.17).

The resin coating was breached, allowing oxygenated water to get below the coating. Thus, differential oxygen concentration cells were established. Rust and other corrosion products exerted pressure on the coating from beneath, causing further delamination and breakage. Such processes are sometimes referred to as "jacking" because of the way in which the oxide "jacks" or lifts the protective layer off the metal surface.

Figure 5.16   Delaminated, phenolic epoxy resin coating on an internal surface. (Magnification: 7.5×.)

Figure 5.17   As in Fig. 5.16 but after a brief chemical cleaning. The coating is more easily visible, and small depressions are visible beneath the coating blisters. (Magnification: 7.5×.)

CASE HISTORY 5.4

Industry:	Primary metals
Specimen Location:	Cooling water return line from a reheat furnace
Specimen Orientation:	Vertical
Environment:	100–200°F (38–93°C); heavily fouled with grease and oil, no chemical treatment other than chlorine-bromine biocide; may blow out lines when completely plugged with grease, oil, and iron
Time in Service:	5 years
Sample Specifications:	⅝ in. (1.6 cm) outer diameter, galvanized carbon steel tubing

Pluggage of cooling water return lines had occurred repeatedly. The material removed from internal surfaces contained large amounts of grease, oil, and rust (Fig. 5.18).

In places, the galvanized layer was consumed, exposing the underlying steel (Figs. 5.19 and 5.20). Wastage was caused by exposure of surfaces to high concentrations of dissolved oxygen. The likelihood of concentration cells operating beneath the heavy deposit and corrosion-product accumulation was high.

Figure 5.18 Heavily fouled and corroded internal surface of a galvanized carbon steel pipe.

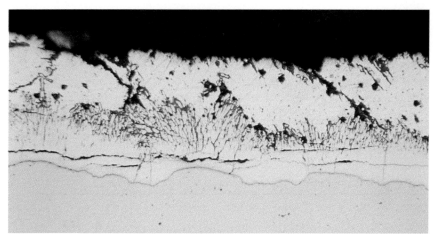

Figure 5.19   As in Fig. 5.18. Typical galvanizing layer on an internal surface in cross section. Compare with Fig. 5.20. The uniform white area at bottom is the tube steel. (Magnification: 220×, unetched.)

Figure 5.20   Small area on an internal surface with a breach in the galvanized layer. (Magnification: 110×, unetched.)

Biologically Influenced Corrosion

General Description

Two basic mechanisms cause biological corrosion. Biologically produced substances may actively or passively cause attack. Each mechanism either accelerates preexisting corrosion or establishes a new form of metal loss. Often the distinction between active and passive attack is vague.

Active biological corrosion may be defined as the direct chemical interaction of organisms with materials to produce new corrosion chemistries and/or the marked acceleration of preexisting corrosion processes. Most of what is currently referred to as *microbiologically influenced corrosion* (MIC) fits into this category. Ultimately, active biological corrosion directly accelerates or establishes new electrochemical corrosion reactions. Factors influencing electrochemistry include pH, area effects, temperature, electrode polarization, flow, oxygen concentration, and electrolyte conductivities. Also, corrosive substances such as acids and ammonia may be formed by biological activity.

In passive attack, biological material acts as a chemically inert substance. Wastage is an indirect consequence of the biological mass or biological by-products. Biomass acts as any deposit accumulation would, providing shielded areas beneath which concentration cells are established. Such masses may also cause flow disturbances, affect heat transfer, and cause erosion (see Case History 11.5 in Chap. 11, "Erosion-Corrosion").

Passive attack involving underdeposit corrosion tends to involve large system surface areas and, hence, accounts for the greatest amount of metal loss, by weight, in cooling water systems. Active attack tends to produce intense localized corrosion and, as such, a greater incidence of perforations.

Active biological corrosion

Most active attack is caused by microorganisms. Potentially troublesome bacteria are either aerobic or anaerobic, although facultative bacteria may grow under either high or low oxygen conditions, changing their metabolic processes and concomittant chemistries accordingly. Furthermore, bacteria grow in a consortium, in which several varieties of organisms coexist in an energy-efficient community. Synergistic effects are common. Biofilm dynamics frequently change with time. One type of microorganism may break down a particular molecule common to the system. A second organism may further degrade the molecular by-products. And a third microorganism may use such by-products to obtain energy.

Each variety of bacteria causes corrosion by creating new reactions occurring at the anode and/or cathode in an electrochemical corrosion cell. Surprisingly, there is considerable disagreement among researchers as to specific reactions involved in microbiologically influenced corrosion. The biological pathways to corrosion can be complex. Such reactions are often open to speculation because of the perceived complexities of microbiological processes. Factors affecting the aggressiveness of corrosive bacteria include temperature, total organic carbon and nitrogen concentrations, flow, oxygen or ammonia concentrations, chemical treatment, pH, and other influences, many of which are unknown.

Four main kinds of bacteria have been linked to accelerated corrosion in cooling water systems:

- Sulfate reducers
- Acid producers
- Metal depositors
- Slime formers

Each classification will be discussed separately.

Sulfate reducers.   The best-known form of microbiologically influenced corrosion involves sulfate-reducing bacteria. Without question, sulfate reducers cause most localized industrial cooling water corrosion associated with bacteria. *Desulfovibrio, Desulfomonas,* and *Desulfotomaculum* are three genera of sulfate-reducing bacteria.

These bacteria are anaerobic. They may survive but not actively grow when exposed to aerobic conditions. They occur in most natural waters including fresh, brackish, and sea water. Most soils and sediments contain sulfate reducers. Sulfate or sulfite must be present for active growth. The bacteria may tolerate temperatures as high as about 176°F (80°C) and a pH from about 5 to 9.

There is disagreement concerning how sulfate reducers cause corrosion. One proposed mechanism suggests that the bacteria cause cathodic depolarization by removing hydrogen from cathodic sites based on undisputed evidence that several sulfate reducers possess enzymes that are capable of converting hydrogen. In this process, inorganic sulfates are reduced to sulfides in the presence of hydrogen. Attack occurs more readily on metal surfaces, probably because of the adsorbed hydrogen forming there. The principal reactions for corrosion of steel are as follows:

Anode:

$$4Fe^0 \rightarrow 4Fe^{++} + 8e^- \qquad (6.1)$$

Water dissociation:

$$8H_2O \rightarrow 8H^+ + 8OH^- \qquad (6.2)$$

Cathode:

$$8H^+ + 8e^- \rightarrow 8H$$
$$\text{(adsorbed on metal surface)} \qquad (6.3)$$

Conversion of sulfate by bacteria:

$$8H \text{ (adsorbed)} + SO_4^{--} \rightarrow S^{--} + 4H_2O \qquad (6.4)$$

Final corrosion products:

$$4Fe^{++} + S^{--} + 6OH^- \rightarrow FeS + 3Fe(OH)_2 \qquad (6.5)$$

Overall reaction:

$$4Fe + SO_4^{--} + 4H_2O \rightarrow 3Fe(OH)_2 + FeS + 2OH^{--} \qquad (6.6)$$

The only above reaction in which bacteria plays a direct role is Reaction 6.4, the cathodic depolarization process.

Acid producers.   Many bacteria produce acids. Acids may be organic or inorganic depending on the specific bacterium. In either case, the acids produced lower the pH, usually accelerating attack. Although many kinds of bacteria may generate acids, *Thiobacillus thiooxidans* and *Clostridium* species have most often been linked to accelerated corrosion on steel.

Thiobacillus thiooxidans is an aerobic organism that oxidizes various sulfur-containing compounds to form sulfuric acid. These bacteria are sometimes found near the tops of tubercles (see Chap. 3, "Tuberculation"). There is a symbiotic relationship between *Thiobacillus* and sulfate reducers; *Thiobacillus* oxidizes sulfide to sulfate, whereas the sulfate reducers convert sulfide to sulfate. It is unclear to what extent *Thiobacillus* directly influences corrosion processes inside tubercles. It is more likely that they indirectly increase corrosion by accelerating sulfate-reducer activity deep in the tubercles.

Clostridia are anaerobic bacteria that can produce organic acids. Short-chain organic acids can be quite aggressive to steel. *Clostridia* are frequently found deep beneath deposit and corrosion-product accumulations near corroding surfaces and within tubercles. Increased acidity directly contributes to wastage.

Metal depositors.   Metal-depositing bacteria oxidize ferrous iron (Fe^{++}) to ferric iron (Fe^{+++}). Ferric hydroxide is the result. Some bacteria oxidize manganese and other metals. *Gallionella* bacteria, in particular, have been associated with the accumulation of iron oxides in tubercles. In fact, up to 90% of the dry weight of the cell mass can be iron hydroxide. These bacteria appear filamentous. The oxide accumulates along very fine "tails" or excretion stalks generated by these organisms.

Evidence of substantial deposition caused by *Gallionella* in tubercles is incomplete, however. The amount of such deposition is usually insignificant compared to the deposit and corrosion-product contribution that occurs in the absence of bacteria.

In some reports *Gallionella* have been associated with manganese and iron deposits that also contain chloride. It has been postulated that deep undercut pits on stainless steels (especially at welds) containing such deposits are indirectly caused by these bacteria, since the iron-manganese deposition can be accelerated by *Gallionella*. In spite of numerous literature citings, however, evidence for stainless steel pitting caused by *Gallionella* is poor. Chlorination also causes oxidation of iron and manganese, and most iron or manganese deposits formed in chlorinated cooling waters contain chloride (see Chap. 4, "Underdeposit Corrosion"), whether iron-oxidizing bacteria are present or not. It is likely that chloride, alone or in conjunction with deposits, is a major cause of observed stainless steel pitting. Sensitization of stainless welds increases all forms of corrosion at these sites.

Other iron oxidizers include *Sphaerotilus, Crenothrix,* and *Leptothrix* species. Each bacterium is filamentous.

Slime formers.   Most slime formers are aerobic, although some, such as *Pseudomonas,* are facultative and can grow in low-oxygen environments. Closed recirculating cooling water systems containing low-oxygen concentrations are usually free of significant slime masses.

Slime layers are a mixture of bacterial secretions called *extracellular polymers,* other metabolic products, bacteria, gases, detritus, and water. Commonly, 99% of the slime layer is water, although much silt and debris may also become entrapped in it.

Slime layers contribute to corrosion both actively and passively. First, since the indigenous slime formers are aerobic, they consume oxygen, stimulating the formation of differential oxygen cells. The slime layer forms an occluding mass, also contributing to differential oxygen cell formation and passive attack (see Chap. 5, "Oxygen Corrosion"). Second, and again because most slime formers are aerobic, these bacteria are present atop corrosion products and deposits in proximity to oxygenated water. Anaerobic bacteria often are found beneath the slime. Sulfate reducers and acid producers frequently occur in high concentrations beneath slime layers.

Other bacteria.   Nitrifying bacteria (principally aerobic) are capable of oxidizing ammonia (NH_3) to nitrate (NO_3^-). The best-known nitrifying bacteria are *Nitrosomonas* and *Nitrobacter.* Biological activity reduces pH, and oxygen concentration falls. A drop in pH is a common sign of nitrifiers. The drop is usually sharp but transient, and corrosion effects are usually slight. Leakage of ammonia into cooling water stimulates growth of *Nitrosomonas.* Hence, ammonia plants often contain such bacteria.

Nitrobacter, an aerobic bacterium, can materially depress pH by oxidizing nitrite (NO_2^-) to nitrate (NO_3^-), in effect producing nitric acid. Acidity may increase until pH is between 3 and 5. Such bacteria require high concentrations of oxygen and cause problems only in oxygenated systems.

Some algae produce dense, fibrous mats in sunlit areas. The mats act as passive corrosion sources. Differential concentration cells may be established. Corrosive anaerobes may grow beneath algal mats. Additionally, high concentrations of dissolved oxygen can be created by vigorously growing algae. In at least one case, dissolved oxygen concentrations measured near cooling water intakes at a steel mill were as high as 10 ppm, well above the calculated oxygen saturation concentration at inlet temperature and pressure. Oxygen concentration fell at night when algae growth stopped. Corrosion rates measured with electric polarization devices followed a similar diurnal cycle.

Passive biological corrosion

Passive corrosion caused by chemically inert substances is the same whether the substance is living or dead. The substance acts as an occluding medium, changes heat conduction, and/or influences flow. Concentration cell corrosion, increased corrosion reaction kinetics, and erosion-corrosion can be caused by biological masses whose metabolic processes do not materially influence corrosion processes. Among these masses are slime layers.

Slime is a network of secreted strands (extracellular polymers) intermixed with bacteria, water, gases, and extraneous matter. Slime layers occlude surfaces—the biological mat tends to form on and stick to surfaces. Surface shielding is further accelerated by the gathering of dirt, silt, sand, and other materials into the layer. Slime layers produce a stagnant zone next to surfaces that retards convective oxygen transport and increases diffusion distances. These properties naturally promote oxygen concentration cell formation.

Slime masses or any biofilm may substantially reduce heat transfer and increase flow resistance. The thermal conductivity of a biofilm and water are identical (Table 6.1). For a 0.004-in. (100-μm)-thick biofilm, the thermal conductivity is only about one-fourth as great as for calcium carbonate and only about half that of analcite. In critical cooling applications such as continuous caster molds and blast furnace tuyéres, decreased thermal conductivity may lead to large transient thermal stresses. Such stresses can produce corrosion-fatigue cracking. Increased scaling and disastrous process failures may also occur if heat transfer is materially reduced.

Metal-reducing bacteria, such as those that convert ferric to ferrous ion, have been suggested as an accelerant for steel corrosion in oxygenated waters. To date, evidence of these bacteria influencing corrosion in industrial systems is scarce.

TABLE 6.1   Thermal Conductivities of Scales and Biofilm

Scale	Thermal conductivity (watt $M^{-1} \, {}^\circ K^{-1}$)
Calcium carbonate	2.26–2.93
Calcium sulfate	2.31
Calcium phosphate	2.60
Magnesium phosphate	2.16
Magnetic iron oxide	2.88
Analcite	1.27
Biofilm	
Water	0.63

SOURCE: N. Zelver, W. G. Characklis, and F. L. Roe, CTI Paper No. TP239A, 1981, Annual Meeting of the Cooling Tower Institute, Houston, Texas.

Certain anaerobic bacteria capable of producing hydrogen may, under special circumstances, contribute to hydrogen embrittlement of some alloys. Once again, if such mechanisms operate, they have very limited applicability in most cooling water systems.

Finally, any living organism dies. Decomposition may generate ammonia at local concentrations high enough to produce stress-corrosion cracking of brass condenser tubes (Fig. 6.1).

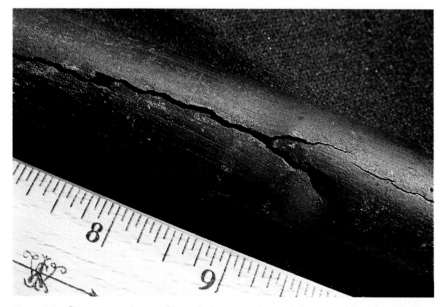

Figure 6.1   Stress-corrosion cracking of a brass condenser tube caused by ammonia from decomposing slime masses lodged on internal surfaces.

Shells, clams, wood fragments, and other biological materials can also produce concentration cell corrosion. Additionally, fragments can lodge in heat exchanger inlets, locally increasing turbulence and erosion-corrosion. If deposits are massive, turbulence, air separation, and associated erosion-corrosion can occur downstream (see Case History 11.5).

Zebra mussels and Asiatic clams are particularly troublesome. Because of their prodigious reproduction and relatively rapid growth rates under favorable conditions, systems such as water intakes, transfer piping, grates, and trash screens can be rapidly plugged. Portions of service water systems in nuclear power plants have been rendered temporarily inoperative due to clams, mussels, and similar organisms. Besides gross occlusion, these organisms increase rates of silt, sand, and detritus accumulation, also accelerating concentration cell corrosion.

Small organisms frequently become embedded within corrosion products and deposits. The organisms may make up a sizable fraction of the deposit and corrosion product. Seed hairs and other small fibers often blow into cooling towers, where they are transported into heat exchangers. The fibers stick to surfaces, acting like sieves by straining particulate matter from the water. Deposit mounds form, reinforced by the fibers (see Case History 11.5).

Locations

Corrosion occurs where biological matter settles on or attaches to surfaces. Attack may also occur away from attached organisms and settled biomass. For example, increased turbulence and air separation downstream of biological deposits may initiate erosion. Biologically produced chemicals such as ammonia, hydrogen sulfide, and acids can increase general corrosion rates in a system, near and away from generating organisms.

Virtually any cooling water system may be affected. Microbiological attack usually occurs where water temperature is below about 180°F (82°C). Slime masses foul water boxes, heat exchanger tubing, and tube sheets. Algal growth covers cooling towers and plugs strainers and water intakes. Fungal growth may cause cooling tower wood rot.

Virtually all metallurgies can be attacked by corrosive bacteria. Cases of titanium corrosion are, however, rare. Copper alloys are not immune to bacterial attack; however, corrosion morphologies on copper alloys are not well defined. Tubercles on carbon steel and common cast irons sometimes contain sulfate-reducing and acid-producing bacteria. Potentially corrosive anaerobic bacteria are often present beneath slime layers and, in the absence of slime or tubercles, in systems containing little oxygen. Oil- and grease-fouled systems generally contain large numbers of bacteria.

Large organisms (those visible to the naked eye) are a sure indicator of bacterial presence and may directly cause attack. These larger organisms often feed off smaller organisms. Clams, mussels, and other shellfish frequently lodge in heat exchanger tube inlets, water intakes, valves, fire protection systems, and service water piping. Lodgment in flowing systems may locally increase turbulence and/or decrease flow. Inlet-end erosion of heat exchanger tubing is a common result.

Critical Factors

Biological matter or offending organisms must be present in a system suffering current attack. However, not every system containing such material or organisms is harmfully affected. A biological presence (current or past) is virtually certain in any cooling water system, but not every cooling water system is significantly attacked.

The presence of biological material is not as critical in making a diagnosis of biologically influence attack as are other factors. *All* the following conditions must exist:

- Biological material must be present at the time of attack.

- Corrosion morphology must be consistent with biological attack.

- Corrosion products and deposits must be characteristic of biological interaction.

Identification

Each form of biologically influenced corrosion can be recognized by examining wastage morphologies, corrosion-product composition and distribution, deposit composition and distribution, biological analysis, and compatible environmental conditions. It should be vigorously stressed that all five factors must be consistent with the diagnosis of biologically influenced corrosion. The presence *alone* of potentially corrosive bacteria or other organisms is not proof of corrosion related to these organisms. Diagnosis is not based on a preponderance of evidence; it is based on *complete consistency* of all evidence.

Active corrosion

Active attack is commonly caused by microorganisms. Four factors must be present for a diagnosis of microbiologically influenced corrosion:

1. Presence of microorganisms or their by-products
2. Microbiologically unique corrosion morphologies
3. Specific corrosion products and deposits
4. Compatible environmental conditions

Each of the above factors is unique to specific bacteria.

Sulfate reducers.   Active sulfate reducers are found in anaerobic environments. These environments may be highly localized, such as inside a tubercle or beneath a spotty deposit. A thin, fairly regular biofilm is difficult to perceive in such microenvironments.

Corrosion-product accumulations containing sulfate-reducer counts of 10^4 or higher colony-forming units per gram are usually associated with significant wastage. Although sulfate reducers probably are not uniformly distributed throughout the deposit and corrosion-product mass (especially in aerated systems), similar counts in large amounts of material taken from corroded steel surfaces are common. Counts in fluids are almost always much lower, but any positive fluid count usually indicates large numbers of viable sessile bacteria somewhere in the system.

Recently, tests have been developed that do not require culturing of sulfate reducers. These tests are based on detecting certain compounds produced by the sulfate reducers and have applicability (in some cases) even if the producing organisms have recently died. Laboratory studies have shown adequate agreement between such tests and live culture analyses when viable organisms are present.

A typical microbiological analysis in a troubled carbon-steel service water system is given in Table 6.2. Table 6.3 shows a similar analysis for a cupronickel utility main condenser that showed no significant corrosion associated with sulfate reducers. When biological counts of sulfate reducers in solid materials scraped from corroded surfaces are more than about 10^4, significant attack is possible. Counts above 10^5 are common only in severely attacked systems.

Planktonic counts (in water samples) are usually unreliable as an indicator of active corrosion. The presence of any sulfate reducers in the water, however, indicates much higher concentrations of these organisms on surfaces somewhere in the system.

Corrosion morphologies.   Sulfate-reducing bacteria frequently cause intense localized attack (Figs. 6.2 through 6.7). Discrete hemispherical depressions form on most alloys, including stainless steels, aluminum, Carpenter 20, and carbon steels. Few cases occur on titanium. Copper alloy attack is not well defined.

TABLE 6.2   Typical Microbiological Analysis in a Service Water
System Pipe Experiencing Microbiologically Influenced Corrosion*

	Water	Deposits and corrosion products
Total aerobic bacteria	490,000	1,100,000
Enterobacter	30	<1000
Pigmented	<70	<1000
Mucoids	<10	<1000
Pseudomonas	100,000	210,000
Spores	12	8700
Total anaerobic bacteria	35	150,000
Sulfate reducers	30	120,000
Clostridia	5	30,000
Total fungi	2	<100
Yeasts	2	<100
Molds	None	<100
Iron-depositing organisms	None	None
Algae		
Filamentous	Very few	None
Nonfilamentous	Few	Few
Other organisms		
Protozoa	Few	None

* All counts expressed as colony-forming units per milliliter or gram.

TABLE 6.3   Typical Microbiological Analysis at Outlet: A Main Condenser
Suffering No Significant Corrosion by Sulfate-Reducing Bacteria*

	Liquid	Deposits
Total aerobic bacteria	310,000	14,000,000
Enterobacter	10	<1000
Pigmented	12,000	<1000
Mucoids	<10	<1000
Pseudomonas	30,000	4,200,000
Spores	None	1000
Total anaerobic bacteria	350	1300
Sulfate reducers	350	1000
Clostridia	None	300
Total fungi	20	1000
Yeasts	None	<100
Molds	20	1000
Iron-depositing organisms	None	None
Algae	None	None
Other organisms	None	None

* All counts expressed as colony-forming units per milliliter or gram.

Figure 6.2 Severely pitted aluminum heat exchanger tube. Pits were caused by sulfate-reducing bacteria beneath a slime layer. The edge of the slime layer is just visible as a ragged border between the light-colored aluminum and the darker, uncoated metal below.

Pit interiors are characteristically smooth and distinctly hemispherical, but become rougher on less-noble alloys. Pits tend to cluster together, overlapping to form irregularly dimpled surfaces. Frequently, a lightly etched aureole surrounds the pit clusters. These etched areas are often produced by shallow corrosion beneath deposit and slime masses that covered the sulfate reducers in service (Figs. 6.3 and 6.4*A* and *B*).

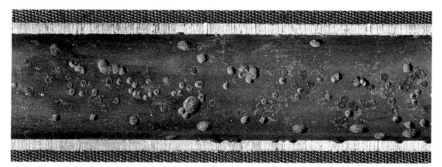

Figure 6.3 Pitting on the waterside surface of a Carpenter 20 heat exchanger tube caused by sulfate reducers.

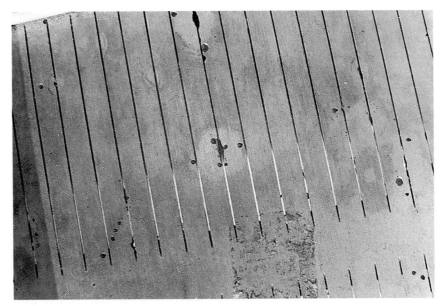

Figure 6.4A   Small pits on a 316 stainless steel plate. A light area covers the clustered pits, marking ghost images of deposit mounds.

Figure 6.4B   As in Fig. 6.4A. Note the smooth hemispherical pit contours and the tendency of the pits to cluster together.

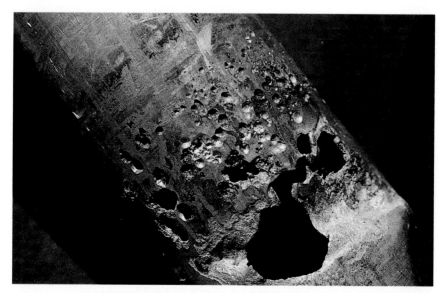

Figure 6.5 Many small hemispherical pits on a 304 stainless steel heat exchanger tube end. The heat exchanger was removed from service and stored vertically for an extended period. Deposit accumulated at the lower tube ends where sulfate reducers flourished.

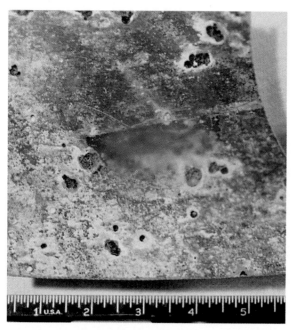

Figure 6.6 Clustered sulfate-reducer pits on a carbon steel tank bottom.

Figure 6.7 A small-diameter carbon steel service water pipe perforated by sulfate-reducer corrosion.

In cross section, pits may exhibit scalloping and mild undercutting, especially on stainless steels (Fig. 6.8). On carbon steels, pit interiors are somewhat less regular. Interiors may contain terrace steps, giving the pits a peculiar bulls-eye pattern when viewed from above. It is tempting to speculate that each terrace marks a corrosion-arrest stage in pit development corresponding to changes in biological activity.

Attack at welds due to bacteria is possible, but it is not nearly so common as is often supposed. Because of residual stresses, microstructural irregularities, compositional variation, and surface irregularities, welds show a predisposition to corrode preferentially by most corrosion mechanisms. Attack is common along incompletely closed weld seams such as at butt welds in light-gauge stainless steel tubing (Fig. 6.9A and B). Attack at carbon steel welds may occur. Figure 6.10 shows a severely corroded carbon steel pipe from a service water sys-

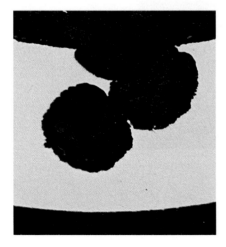

Figure 6.8 Scalloped, partially undercut pits in a cross section of a 316 stainless steel tube.

tem. Wastage began at a circumferential weld employing a backing ring (see Case History 6.1).

Corrosion products and deposits. All sulfate reducers produce metal sulfides as corrosion products. Sulfide usually lines pits or is entrapped in material just above the pit surface. When freshly corroded surfaces are exposed to hydrochloric acid, the rotten-egg odor of hydrogen sulfide is easily detected. Rapid, spontaneous decomposition of metal sulfides occurs after sample removal, as water vapor in the air adsorbs onto metal surfaces and reacts with the metal sulfide. The metal sulfides are slowly converted to hydrogen sulfide gas, eventually removing all traces of sulfide (Fig. 6.11). Therefore, only freshly corroded surfaces contain appreciable sulfide. More sensitive spot tests using sodium azide are often successful at detecting metal sulfides at very low concentrations on surfaces.

Figure 6.9A Irregular deposit and corrosion-product mounds containing concentrations of sulfate-reducing bacteria on the internal surface of a 316 stainless steel transfer line carrying a starch-clay mixture used to coat paper material. Attack only occurred along incompletely closed weld seams, with many perforations. Note the heat tint, partially obscured by the deposit mounds, along the circumferential weld.

Figure 6.9*B* Longitudinal cross section showing a pit tunneling into an incompletely closed weld seam.

Sulfides are intermixed with iron oxides and hydroxides on carbon steels and cast irons. The oxides are also produced in the corrosion process (Reaction 6.6). Although theoretical stoichiometry of 1 to 3 is often suggested between sulfide and ferrous hydroxide, empirically the ratio of iron sulfide to ferrous hydroxide is highly variable. Sulfide decomposes spontaneously upon exposure to moist air. Additionally, corrosion-product stratification is marked, with sulfide concentration being highest near metal surfaces.

Figure 6.10 A perforated carbon steel pipe at a weld-backing ring. The gaping pit was caused by sulfate-reducing bacteria (see Case History 6.1).

Figure 6.11   A ⅟₁₆-in. (0.16-cm) diameter pit on the internal surface of a 316 stainless steel tube. Note the absence of significant corrosion products or deposits in the pit.

Stainless steels attacked by sulfate reducers show well-defined pits containing relatively little deposit and corrosion product. On freshly corroded surfaces, however, black metal sulfides are present within pits. Rust stains may surround pits or form streaks running in the direction of gravity or flow from attack sites. Carbon steel pits are usually capped with voluminous, brown friable rust mounds, sometimes containing black iron sulfide plugs (Fig. 6.10).

Acid producers.   Corrosion usually is moderate and localized. Almost all significant attack is associated with anaerobic bacteria (facultative and obligate), as aerobic acid-producing varieties usually reside near the top of deposits and corrosion products contacting oxygenated waters. Thus, the direct effect on corrosion at metal surfaces is limited. Additionally, although acidic products may be expected to increase corrosion rates, acidity cannot be pronounced in deposits; to put it simply, the deposits and corrosion products would dissolve at sufficiently acidic pH.

Clostridia frequently are found where sulfate-reducing bacteria are present, often in high numbers inside tubercles. A typical microbiological analysis of tubercular material removed from a troubled service water system main is given in Table 6.4. *Clostridia* counts above 10^3/g of material are high enough to cause concern. When acid producers markedly accelerate corrosion, counts are often greater than sulfate-reducer counts.

Surfaces beneath affected tubercles often have a striated contour due to increased acidity (see Fig. 3.24). Striated surfaces are caused by preferential attack along microstructural and microcompositional irregularities that have been elongated during steel rolling (see Chap. 7, "Acid Corrosion").

Short-chain and low molecular weight organic acids, such as acetic acid and formic acid, can be formed by certain bacteria. The resulting organic acid salts are not easily detected without specialized analytical equipment in a laboratory.

Passive corrosion

Wastage is caused by biological material accumulating on or near surfaces. Microbial mats such as slime layers and massive accumulations of larger organisms cause most damage.

Slime layers.   Sliming is easily recognized by the gelatinous mass that accumulates on surfaces. Heat exchanger tubes feel greasy or slick to the touch. Tube sheets and headboxes are frequently affected (Fig. 6.12*A* and *B*). Slime may be colored by dirt and other debris that accumulates in the gooey mass. When dry, slime forms a thin, brittle layer that exfoliates in large patches (Fig. 6.13).

TABLE 6.4   **Typical Microbiological Analysis of Tubercular Material Containing High Concentrations of *Clostridia* ***

Total aerobic bacteria	1,300,000
Enterobacter	<1000
Pigmented	<1000
Mucoids	<1000
Pseudomonas	150,000
Spores	900
Total anaerobic bacteria	24,400
Sulfate reducers	22,200
Clostridia	2200
Total fungi	<100
Yeasts	<100
Molds	<100
Iron-depositing organisms	None
Algae	None
Other organisms	None

* All counts expressed as colony-forming units per gram.

Figure 6.12*A*   A gelatinous slime layer on a heat exchanger tube sheet.

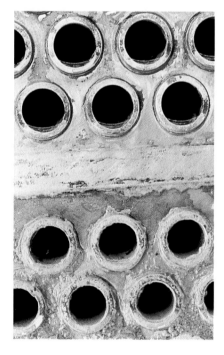

Figure 6.12*B*   As in Fig. 6.12*A*, but split above and below the water box to show the clean outlet sheet (top) and the slimed inlet sheet (bottom).

Figure 6.13 A dried slime layer peeling off water box surfaces on a small heat exchanger.

Sliming is often more severe on inlet tube sheets than on outlets. Bacterial counts are usually quite high, exceeding tens of millions in most slime layers (Table 6.5). Sulfate-reducer counts are also usually high. In waters taken from such systems, bacteria counts are often several orders of magnitude lower.

Passive attack beneath slime is usually general, and rusting on steel may color surfaces brown and red (Fig. 6.14). If sulfate reducers are present, pitting may be superimposed on the generally corroded surface (see Fig. 6.2).

TABLE 6.5 Typical Microbiological Evaluation of Slime Layer in Fig. 6.13 and Typical Counts in Cooling Water Contacting Heavily Slimed Surfaces*

	Slime layer	Water
Total aerobic bacteria	46,000,000	8000
Enterobacter	100,000	<1000
Pigmented	None	<1000
Mucoids	None	<1000
Pseudomonas	9,100,000	<1000
Spores	9200	None
Total anaerobic bacteria	33,000	<10
Sulfate reducers	30,000	None
Clostridia	3000	None
Total fungi	3100	1000
Yeasts	<100	<10
Molds	3000	<10
Iron-depositing organisms	None	None
Algae	None	None
Other organisms	None	None

* All counts expressed as colony-forming units per milliliter or gram.

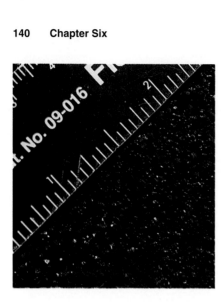

Figure 6.14 A generally corroded carbon steel surface after the slime layer was removed.

Rolling oil systems often experience etching beneath deposit accumulations containing large numbers of bacteria. In steel rolling oil systems when aerobic counts in the mill coolant reach about 2×10^7, noticeable degradation of the coolant occurs with a significant drop in pH due to microbiological breakdown of oil components to low molecular weight organic acids. Also, fungi counts exceed 10^3. It is supposed that fungal growth is stimulated by the presence of large masses of dead bacteria.

Large organisms.   A great variety of large organisms foul cooling water systems. Clams and mussels are frequently found in cooling tower basins, condensers, heat exchangers, and water supply systems. Many creatures find conditions very favorable for growth in these locations. Small organisms become incorporated into deposits and corrosion products. Vegetative fibers, insects, and other biological materials are blown into cooling towers and make their way throughout the cooling system.

Clams, mussels, and other bivalves are a serious problem in many cooling water systems. Zebra mussels and freshwater Asiatic clams are relative newcomers to the United States. Because of a lack of natural predators and prodigious reproduction rates, they have rapidly become a serious threat to the operation of many cooling water systems.

Shells from similar animals have fouled many cooling water systems on seacoasts for some time (Figs. 6.15 and 6.16). Shells tend to lodge near tube inlet ends in shell-and-tube heat exchangers; objects slightly smaller are swept through tubes, and larger ones cannot fit into openings. Brass and cupronickel tubing may suffer severe, localized

Figure 6.15   A main mill water-strainer basket caked with clams.

erosion-corrosion where shells touch surfaces (see Chap. 11, "Erosion-Corrosion"). Occasionally, multiple erosion sites will be present in a single tube, marking shell arrest points as the shell "walks" down the tube, becoming briefly wedged at various locations. When erosion-corrosion occurs due to lodged objects, there will always be at least two

Figure 6.16   Small clams lodged in heat exchanger tube inlet ends. (*Courtesy of Rick Ruckstuhl, Nalco Chemical Company.*)

individual sites of attack corresponding to simultaneous points of contact with the lodged object.

Massive accumulations of shells may obstruct pipes, screens, valves, and other water passages. Often obstruction occurs immediately after or during high-flow conditions in normally stagnant systems. Large numbers of weakly adhering and/or dead shells are detached from surfaces during high flow, causing valve jamming and system pluggage.

Small organisms.   Admittedly, the distinction between large and small organisms is somewhat arbitrary—all large organisms must have been small at one time. From an empirical viewpoint, however, small organisms may be loosely defined as those creatures less than 0.4 in. (1 cm) in length. Into this category falls a wide variety of invertebrates and plants. In a surprising number of cases, substantial amounts of deposits are composed of small organisms (Fig. 6.17A and B). Diatoms, larval bivalves, and other small organisms are frequently found attached to surfaces in cooling water systems (Figs. 6.18 through 6.20A and B). Small organisms tend to accelerate deposition. If fibrous, organisms may act as sieves and/or become encased in particulate.

Figure 6.17A   A corrosion-product and deposit mound on a mild steel service water pipe honeycombed by small tubelike organisms. Each hole is approximately 0.01 in. (0.025 cm) in diameter.

Figure 6.17B   Close-up of small tubelike "casts." More than 50% by volume of the surrounding material is composed of these casts.

Figure 6.18   Pinhead-sized, shell-like organisms covering a carbon steel pipe surface.

Figure 6.19 Small organisms found inside a tubercle. Each is about 0.05 in. (0.13 cm) high. The organisms have segmented, fibrous stalks and bivariate heads.

In many cases small organisms cause manganese, iron, silicate, and calcium deposit enrichment. These organisms can be seen upon close inspection of affected surfaces. However, many small organisms without exoskeletons shrivel upon drying. Only creatures containing substantial inorganic constituents and other "skeletal" remnants not subject to degradation by dessication are easily visible in dried material. Often organisms are encased or are cloaked by deposit and corrosion products, making visual recognition difficult.

Figure 6.20A A 1 in. (2.5 cm) diameter mild steel service water pipe containing many small cone-shaped organisms.

Figure 6.20B   Close-up of organisms shown in Fig. 6.20A.

Elimination

Biological corrosion and deposition may be prevented by chemical treatment, system operation, and system design. Economics alone often favor chemical treatment. However, costs can usually be further reduced by appropriate system design and operation. Water treatment using chlorine, bromine, ozone, or other chemicals can control almost any biological problem. However, discharge limitations, associated corrosion, and other problems often restrict chemical use. Shocking with massive amounts of biocides may be effective in treating some systems, but not all systems will respond identically. Shocking heavily fouled systems may produce sloughing of large biological mats that plug components. After shocking, bacterial growth may be rapid, and the system can return to its previous state quickly. It is imperative that biological control not be erratic. It is much easier and decidedly less costly to maintain good control than to bring a seriously troubled system back into control.

Perhaps the single greatest enemy of good biological control in cooling water systems is low flow. Stagnant regions are almost always the first places in which biologically influenced corrosion occurs. Dead legs should be eliminated. Flow not only prevents settling of detritus but also replenishes biocides and corrosion inhibitors.

Dissolved oxygen is the enemy of anaerobic bacteria. Naturally occurring corrosion consumes oxygen. Without convective replenishment, oxygen-depleted zones form and corrosion by anaerobes is usually worse.

Large organisms such as clams and mussels can be removed by strainers and grates. Bivalves are relatively resistant to some biocides but will succumb if treatment is persistent. Larval forms must be eliminated by chemical treatment. Knowledge of life cycles and related information can be necessary in controlling large organisms. Treatment is much more effective at certain times in each organism's development.

Cautions

The importance of biologically influenced corrosion (particularly microbiologically influenced corrosion) has been underestimated for many years. Recently, more attention has been paid to biological forms of corrosion—yet more attention does not always mean an improved situation.

Many cases of "biologically influenced corrosion" are still being misdiagnosed. Microorganisms are ubiquitous. Bacteria are present in virtually all cooling water systems. Indeed, corroded systems usually contain a diverse group of microorganisms (as do unattacked systems). The critical issue to address when investigating any case of suspected microbiologically influenced corrosion is the extent to which organisms have influenced attack. Is the corrosion exclusively associated with biological activity, or would essentially the same corrosion have occurred in the absence of organisms? Do other coexisting or competing forms of corrosion better represent observed damage morphologies, deposits, and corrosion products? Such questions can be answered but only by careful, unprejudiced, critical examination by informed investigators well versed in industrial failure analysis.

Many forms of corrosion resemble each other. However, it is just as true that each form of corrosion produces a unique "fingerprint" by which it can be differentiated from all other forms of attack. Biologically influenced corrosion is no exception.

The presence of organisms (large or small) in proximity to corrosion by *itself* is not proof of biologically influenced corrosion, any more than a correlation of lunar phases with stock market fluctuations establishes a lunar-financial connection. It should be stressed vigorously that all evidence must be consistent with any single corrosion mode before a definitive diagnosis can be made (see "Critical Factors" above). Further, all alternative explanations must be carefully examined.

Related Problems

See also Chap. 3, "Tuberculation"; and Chap. 4, "Underdeposit Corrosion."

CASE HISTORY 6.1

Industry:	Nuclear utility
Specimen Location:	Emergency service cooling water system
Specimen Orientation:	Horizontal
Environment:	Internal: Cooling water near ambient temperature, intermittent low flow, daily chlorination for 1 hour
	External: Ambient air
Time in Service:	5 years
Sample Specifications:	3.5 in. (8.9 cm) outer diameter, carbon steel pipe, wall thickness ~0.225 in. (0.572 cm)

A weeping leak developed in a carbon steel emergency service water pipe at a circumferential weld employing a weld-backing ring. A rubberized saddle clamp was used to plug the leak temporarily. After several weeks the section was cut out of the system and the failure was examined.

Internal surfaces were heavily tuberculated. The circumferential weld-backing ring was heavily corroded (Fig. 6.21). The perforation shown in Fig. 6.10 occurred at the ring, which was entirely consumed by corrosion in places. At intact areas the weld was sound, showing adequate penetration and no unusual characteristics.

The weld was riddled with mildly undercut, gaping pits. Attack was confined to fused and heat-affected zones, with a pronounced lateral or circumferential propagation (as in Fig. 6.10). The resulting perforation at the external surface was quite small. Pits were filled with deposits, friable oxides, and other corrosion products. Black plugs embedded in material filling the gaping pit contained high concentrations of iron sulfide. Bulk deposits contained about 90% iron oxide. Carbonaceous material was not detected.

The tube was received while still wet, less than 48 hours after removal from the system. Water samples were also taken near the failure site. Resulting microbiological analyses are given in Table 6.6.

The large amounts of sulfide, high bacterial counts, pit morphology, and other factors strongly indicate corrosion was accelerated by sulfate-reducing bacteria. Microscopic evidence also implicated *Clostridia*. The crevice between the backing ring and the tube wall provided a shielded location for deposit concentration and created an environment favorable to anaerobes. Residual weld stress and microstructural irregularities further promoted attack.

Pit morphology is particularly interesting. The major depression is cavernous and shows lateral propagation. However, only a small weeping perforation is present. It is tempting to speculate that once the pipe wall was breached, air in the vicinity of the perforation limited anaerobic activity, thus producing the laterally propagating pit.

Figure 6.21   A longitudinally split carbon steel emergency service water pipe, as in Fig. 6.10. Note the circumferential weld-backing ring, which is consumed on one side.

TABLE 6.6   Microbiological Evaluations of Water and Deposits in Case History 6.1*

	Water	Deposits and corrosion products
Total aerobic bacteria	2,400,000	1,700,000
Enterobacter	<1000	<1000
Pigmented	<10,000	<10,000
Mucoids	<10,000	<10,000
Pseudomonas	20,000	150,000
Spores	2000	9500
Total anaerobic bacteria	35,000	32,000
Sulfate reducers	30,000	25,000
Clostridia	5000	7000
Total fungi	<100	300
Yeasts	<100	<100
Molds	<100	300
Iron-depositing organisms	None	None
Algae	None	None
Other organisms	None	None

* All counts expressed as colony-forming units per milliliter or gram.

CASE HISTORY 6.2

Industry:	Steel
Specimen Location:	Cooling water supply line thermocouple housing
Specimen Orientation:	Vertical
Environment:	90–110°F (32–43°C), pH ≈ 8.5, flow 5–6 ft/s (1.4–1.8 m/s), conductivity 60 µmhos
Time in Service:	~1 year
Sample Specifications:	3.5 in. (8.9 cm) long, 0.5 in. (1.3 cm) diameter, mild steel, wall thickness 0.12 in. (0.30 cm)

Surfaces exposed to water contained many localized areas of wastage (Fig. 6.22). Corroded areas were covered by brittle black or brown caps over irregular metal-loss regions. Many small fibers were embedded in corrosion products, and many more were wrapped around surfaces. Most fibers were about 0.001 in. (0.003 cm) in diameter and were hollow. Surface analysis showed that up to 20% by weight of the fibers was sodium chloride.

The fibers were vegetative in origin, possibly seed hairs. They concentrated salt from the brackish water in which their parent plant grew. These fibers blew into cooling towers and were circulated throughout the cooling water system. Some became attached to rough surfaces. The high concentration of salt in the fibers caused an increased localized corrosion rate at deposit points.

Figure 6.22 A thermocouple housing showing localized wastage. Small fibers containing high concentrations of salt were found in corroded areas.

CASE HISTORY 6.3

Industry:	Alcohol production
Specimen Location:	Condenser
Specimen Orientation:	Horizontal
Environment:	Internal: Cooling water exit temperature 120°F (29°C), calcium 270 ppm, magnesium 150 ppm, sodium 94 ppm, manganese 1.2 ppm, chloride 94 ppm, fluoride 9.3 ppm, pH 7.3 ppm
	External: Ethyl alcohol
Time in Service:	18 months
Sample Specifications:	0.75 in. (1.9 cm) outer diameter, 304 stainless steel tubing

The section was perforated in several locations due to severe, localized wastage on internal surfaces (Fig. 6.23*A* and *B*). The cooling water had a history of low-pH excursions, with documented depressions to a pH below 5. The system also had been plagued with high sulfate-reducing bacteria counts.

Internal surfaces were covered with brown and tan deposits and corrosion products; pits were present beneath small mounds of reddish-brown oxide (Fig. 6.23A). When deposits and corrosion products were removed, many pits were revealed (Fig. 6.23B). Most pits were shallow, but some were deep and undercut. The deepest pits had distinctly hemispherical contours (Fig. 6.23C). In places, smaller undercut pits were found growing off the larger hemispherical pits. Chemical spot tests revealed large concentrations of sulfides in pits.

Pitting had two distinct causes. Sulfate reducers had formed the large hemispherical pits. The more undercut pits were formed during a low-pH excursion involving mineral acid after the sulfate-reducing bacteria became inactive. It is likely the low-pH excursion deepened preexisting sulfate-reducer pits, causing final perforation.

Figure 6.23A   A longitudinally split 304 stainless steel condenser tube covered with deposits.

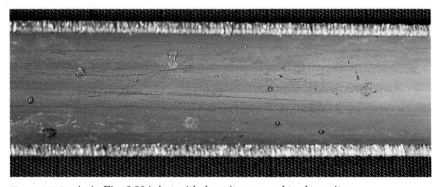

Figure 6.23B   As in Fig. 6.23A, but with deposits removed to show pits.

Figure 6.23C   Pits in profile. Note the smaller pit growing off the large hemispherical pit above.

CASE HISTORY **6.4**

Industry:	Utility
Specimen Location:	Main condenser
Specimen Orientation:	Horizontal
Environment:	Internal: Brackish water, chloride >10,000 ppm, 65–100°F (18–38°C), pH 8.0–8.5, flow 7 ft/s (2.1 m/s)
	External: Condensing steam
Time in Service:	2 years
Sample Specifications:	⅞ in. (2.2 cm) outer diameter, 90:10 cupronickel tubing

After only 4 months of service, the main condenser at a large fossil utility began to perforate. Initial perforations were due to erosion-corrosion (see Case History 11.5). Small clumps of seed hairs entering the condenser after being blown into the cooling tower were caught on surfaces. The entrapped seed hairs acted as sieves, filtering out small silt and sand particles to form lumps of deposit (Fig. 6.24A and B). Immediately downstream from each deposit mound, an erosion-corrosion pit was found.

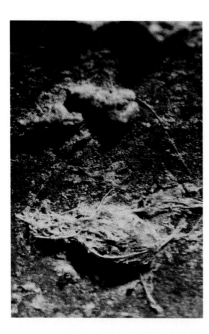

Figure 6.24*A*   Small mounds of deposit entrapped by seed hairs that stuck to surfaces.

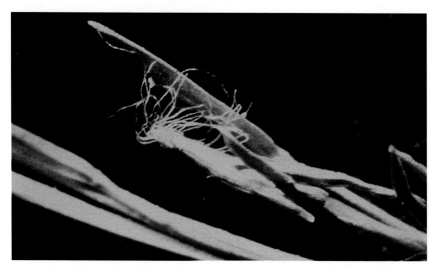

Figure 6.24*B*   The source of the seed hairs in Fig. 6.24*A*. These plants grew immediately around the cooling tower. The white fibers are seed hairs.

CASE HISTORY 6.5

Industry:	Nuclear utility
Specimen Location:	Emergency service water system piping to reactor core spray
Specimen Orientation:	Vertical and horizontal (90° bend)
Environment:	Internal: Ambient river water, unclarified
	External: Ambient air
Time in Service:	~10 years
Sample Specifications:	1 in. (2.5 cm) outer diameter, mild steel pipe

After almost 10 years of service, several pipes were removed because of restricted flow associated with accumulation of corrosion products and macrofouling (Figs. 6.17 and 6.20). The most heavily fouled sections contained unclarified river water and remained stagnant for as long as 30 days at a time. Eighteen months before removal of the pipes, parts of the system had been cleaned using high-pressure water jets. No failure had occurred. Wastage was nowhere deeper than about 0.030 in. (0.076 cm). However, small-diameter tubes were severely obstructed. Up to half the volume of obstructing material consisted of small tubelike organisms (Fig. 6.17).

Chemical analysis showed that each organism contained up to 50% silica by weight. Each was coated with iron oxides, silt, and other deposits and corrosion products. In places, large deposit accumulations were clearly correlated with large numbers of organisms.

From the orientation of the tubelike organisms it was clear that they were flourishing even though the system was periodically chlorinated (Fig. 6.25). Clearly, the biocidal treatment was inadequate.

Figure 6.25 Low-power scanning electron micrograph of small tubelike organisms in Figs. 6.17 and 6.20. Note how all organisms (except one) point into the direction of flow.

CASE HISTORY 6.6

Industry:	Nuclear utility
Specimen Location:	Service water system, branch line
Specimen Orientation:	Horizontal
Environment:	Internal: Surface near ambient temperature, daily chlorination for 2 hours
Time in Service:	7 years
Sample Specifications:	4 in. (10 cm) inner diameter, schedule-40 carbon steel pipe

Many pipes were permitted only a fraction of their normal flow. Close visual inspection of surfaces revealed many small organisms lying atop and beneath tubercles (Figs. 6.18 and 6.19). Surfaces were analyzed using energy-dispersive spectroscopy. Up to 40% of the material covering small, shell-like organisms was manganese. Calcium concentration was as high as 15%, and sulfur concentrations approached 10%.

A perforation in similar piping was caused by wastage beneath a tubercle. Although potentially aggressive bacteria were present in this system, no evidence of microbiologically influenced corrosion was found on this section.

On-site inspection of regions near the emergency service water intake structures revealed loss of up to 40% of the nominal pipe-wall thickness. Deposits were dried, ignited, and analyzed. A 1% solution of dried ash had a pH between 4.4 and 5.5. Sulfur concentrations were as high as 7.6%. By elemental analysis, it was determined that sulfur was concentrated near large organisms and was almost absent in corrosion products farther away.

The above analysis indicates that the high concentrations of sulfur-containing deposits and corrosion products were caused by the influence of large organisms. Bacterial contributions to corrosion and associated fouling were minimal.

CASE HISTORY 6.7

Industry:	Primary metals
Specimen Location:	Rolling oil coolant tanks, cold mill
Specimen Orientation:	Vertical wall
Environment:	Rolling oil emulsion, 125°F (52°C)
Time in Service:	2 years
Sample Specifications:	Carbon steel plate

Rolling oil tanks were corroded on surfaces contacting the emulsion. Small pitlike depressions were present beneath aluminum soap deposits. Each pit was surrounded by a lightly etched region exactly mirroring deposit patterns (Fig. 6.26).

Organic acids concentrated in deposits and caused most attack. However, the peculiar attack morphology of a deep, localized area of corrosion surrounded by lightly etched areas was not characteristic of acid corrosion.

Microbiological analysis of material removed from corroded areas showed high aerobic and relatively high anaerobic counts (Table 6.7). Fungi counts were also elevated, indicating relatively high die-off of aerobic bacteria; that is, the coolant was approaching its useful life.

This case history illustrates the paradox so often encountered in microbiologically influenced corrosion. Clearly, two corrosion mechanisms were operating in the system, namely, acid attack and microbiologically influenced corrosion. To what degree each mechanism contributed to wastage is difficult to quantify after the fact. This was especially the case here, since other areas of the rolling oil system were attacked by a predominantly acidic form of corrosion.

Figure 6.26 Small, pitlike depressions near the center of lightly etched regions. Each lightly etched region was covered by a gelatinous soap deposit in service. Note how the deeper, more localized metal loss occurred away from the lightly etched border.

**TABLE 6.7   Microbiological Evaluation of Material
Removed from Corroded Areas in Case History 6.7***

Total aerobic bacteria	18,000,000
Enterobacter	<1000
Pigmented	11,000,000
Mucoids	1,000,000
Pseudomonas	4,000,000
Spores	500
Total anaerobic bacteria	2000
Sulfate reducers	2000
Clostridia	<1000
Total fungi	46,000
Yeasts	5000
Molds	41,000
Iron-depositing organisms	None
Algae	None
Other organisms	None

* All counts expressed as colony-forming units per gram.

Acid Corrosion

General Description

Acid corrosion in cooling water systems is usually caused by an upset. Acid spills, process leaks, airborne gas contamination, and unintentional overfeed of acid during chemical water treatment are the major causes of attack. Chronic, long-lasting attack can occur if pH does not fall to extremely low values. Metal loss caused by acid conditions can be startlingly high, and the severity of attack quickly becomes apparent through rapid failure. Laboratory corrosion studies of 316 stainless steels exposed to a 1% solution of highly aerated and rapidly stirred hydrochloric acid at 149°F (60°C) produced short-term corrosion rates of 81,400 mil/y (208 cm/y); that is, 0.22 in. (0.56 cm) per day.

Many factors influence acid corrosion. Metallurgy, temperature, water turbulence, surface geometry, dissolved oxygen concentration, metal-ion concentration, surface fouling, corrosion-product formation, chemical treatment, and, of course, the kind of acid (oxidizing or nonoxidizing, strong or weak) may markedly alter corrosion.

Steel

The corrosion rate of carbon steel (and most cast irons) in pure water is about constant between a pH of 4 and 10 (see Fig. 5.5). In solutions containing strong acids such as hydrochloric and sulfuric, normally protective iron oxides are dissolved below a pH of 4. Dissolution of the oxide layer occurs at slightly higher pH (5 or 6) in weaker acids such as carbonic and a variety of organics. This seeming paradox is explained by an analysis of what is meant by weak and strong acids.

Acid strength depends on the tendency of the acid to dissociate into a hydrogen ion and counter ion. Weak acids, such as carbonic and organics, dissociate only slightly compared to strong mineral acids such as sulfuric, hydrochloric, and nitric. Since dissociation is incomplete in weak acids at a constant pH, more hydrogen ion is available to attack the iron oxide. Hydrogen ions consumed in corrosion cannot be replaced in strong acid solutions since dissociation of these acids is almost complete. In weak acid solutions, however, more hydrogen ions can be produced by dissociation as the hydrogen ions in solution are consumed during corrosion.

The main anodic reaction in acid solutions is given in Reaction 7.1; iron is dissolved at exactly the rate of the cathodic process:

Anode: $$Fe \rightarrow Fe^{++} + 2e^{-} \qquad (7.1)$$

Cathode: $$2H^{+} + 2e^{-} \rightarrow H_2 \uparrow \qquad (7.2)$$

Thus, hydrogen gas is generated in most acid solutions. If dissolved oxygen is present, the cathodic Reaction 7.2 can be accelerated by depolarization as in Reaction 7.3:

Cathode depolarization: $$4H^{+} + O_2 + 4e^{-} \rightarrow 2H_2O \qquad (7.3)$$

Depolarization effects are important in weak and nonoxidizing acids. The synergistic effects of oxygen with carbonic acid and a variety of organic acids are well known. Oxygen effects are largest for weak and dilute inorganic acids (Table 7.1). When acids are concentrated, hydrogen evolution is so great that oxygen cannot easily reach the corroding surface. Oxidizing acids such as nitric acid show virtually no synergism with oxygen, since they act as depolarizers. Thus, the addition of oxygen has little or no effect on corrosion rates in oxidizing acids.

Access of oxygen to steel surfaces during corrosion influences the wastage process in nonoxidizing acids. Fluid velocity can influence the amount of oxygen reaching the metal surface and, therefore, the corrosion rate. In deaerated acid solutions, steel corrosion rate is constant with fluid velocity. If dissolved oxygen is present, however, the corrosion rate is proportional to fluid velocity.

During acid cleaning or severe acid upsets, ferric-ion concentration may increase, albeit much more slowly than ferrous-ion concentration, to high levels. Resulting corrosion can be severe. Iron is oxidized, and ferric ions reduced, as in Reaction 7.4, on bare steel surfaces that develop during system cleaning or upset:

$$Fe° + 2Fe^{+++} \rightarrow 3Fe^{++} \tag{7.4}$$

Carbonic acid is corrosive to steel even if oxygen is absent. Corrosion is accelerated greatly by the presence of dissolved oxygen, however.

Stainless steel

Acidic attack on stainless steels differs from corrosion on nonstainless steels in two important respects. First, nonoxidizing acid corrosion is usually more severe in deaerated solutions; second, oxidizing acids attack stainless steel far less strongly than carbon steel. Hence, nitric acid solutions at low temperatures cause only superficial damage, but hydrochloric acid causes truly catastrophic damage.

Stainless steels tend to pit in acid solutions. Pits form local areas of metal loss associated with breakdown of a protective oxide layer. Breakdown is stimulated by low pH as well as by the decrease of dissolved oxygen in occluded regions. Small, active pit sites form and remain stable because of the large ratio of cathodic surface area (unattacked metal surface) to the pit area. Active corrosion in the pit cathodically protects immediately adjacent areas. If conditions become very severe, pitting will give way to general attack as more and more of the surface becomes actively involved in corrosion.

Copper alloys

Copper corrosion by mineral and organic acids is controlled to a large degree by the presence of oxidizing agents. Nonoxidizing acids such as organic acids cause very little corrosion if oxygen concentration is very low. Hence, just above boiling temperatures copper is virtually impervious to attack. If cupric or other metallic salts that might be reduced

TABLE 7.1   Effect of Dissolved Oxygen on Corrosion of Mild Steel in Acids Corrosion Rate (in./y)

Acid	Under O_2	Under H_2	Ratio
6% acetic	0.55	0.006	87
6% H_2SO_4	0.36	0.03	12
4% HCl	0.48	0.031	16
0.04% HCl	0.39	0.0055	71
1.2% HNO_3	1.82	1.57	1.2

SOURCE: L. W. Whitman and R. Russell, *Industrial & Engineering Chemistry* 17, p. 348 (1925).

are present, attack may be severe, even in the absence of dissolved oxygen. Oxidizing acids such as nitric acid cause severe attack. Zinc brasses and cupronickels behave in much the same way as copper.

The corrosion of copper by carbonic acid deserves special attention. There is a synergism between oxygen and carbonic acid with regard to corrosion. Carbonic acid in the absence of oxygen is not corrosive to most copper alloys. However, corrosivity can be appreciable if oxygen is present.

Aluminum

Aluminum corrodes at a fairly low rate between a pH of 5.5 and 8.5 at room temperature. At concentrations between 50% and 95%, sulfuric acid causes rapid attack; below 10%, corrosion is much less. Hydrochloric acid is quite corrosive in all but dilute concentrations. The corrosion rate in hydrochloric acid increases 100-fold as temperature increases from 50°F (10°C) to 176°F (80°C) in a 10% hydrochloric acid solution.

Aluminum is resistant to nitric acid at concentrations above 80%. At 50% nitric acid concentration at room temperature, corrosion rates are as high as 0.08 in. (0.20 cm) per year.

Phosphoric acid can be used to etch aluminum uniformly at dilute concentrations. At higher concentrations, attack is rapid.

Organic acids—except formic, oxalic, and some chlorine-containing acids—do not appreciably attack aluminum near room temperature. In most acids, the corrosion rate increases slightly with flow velocity.

Locations

Virtually any cooling water component contacting low-pH water may be corroded. However, there are appreciable differences as to corrosion severity and initiation times depending on alloy composition, kind of acid, and location.

In general, the higher the residual or applied metal stress, the more severe the corrosion at a given acidic pH. This explains why many heat exchanger tube ends are often attacked so severely (Fig. 7.1). Tube ends that have been rolled or welded often contain high residual stress. Further, crevices are sometimes present in which acidic species may concentrate (see Chap. 2, "Crevice Corrosion"). Screens, rolled sheet metal, and other highly worked metals (not stress relieved) are also prone to attack.

Alloys whose corrosion resistance depends on forming a protective oxide layer, such as stainless steel, are susceptible to severe localized attack when pH falls as a result of nonoxidizing acid excursions. However, there is usually a period of immunity from attack corresponding to the time necessary to breach the protective oxide layer. Hence, stainless steels usually experience very little attack during brief low-pH excursions. During long acidic exposure, however, pits may begin and grow at an ever-increasing rate. Surfaces containing pits, crevices, and other shielded areas are more likely to be attacked by low-pH conditions. Occluded sites allow concentration of acidic materials and assist the so-called *autocatalytic* pitting process.

Corrosion involving nonoxidizing acids can be highly sensitive to flow. Thus regions of high flow and turbulence are often more severely attacked than more quiescent regions. Weirs, lips, and other flow obstructions increase turbulence and thus corrosion. Pipe elbows, tees, and joints are frequently attacked. Outer curves at pipe bends often are more severely wasted than inner bends.

Dissolved oxygen, water, acid, and metal-ion concentrations can have a pronounced effect on acid corrosion. For example, copper is vigorously attacked by acetic acid at low temperatures; at temperatures above boiling, no attack occurs because no dissolved oxygen is present.

Figure 7.1   Brass condenser tube inlet end severely corroded by an acid upset involving sulfuric acid.

Hence, copper heat exchanger tubes handling acetic acid can be more seriously corroded at low temperatures than at high temperatures. Sulfuric acid at room temperature is handled routinely in carbon steel drums and tanks when water concentration is low, but it becomes extremely corrosive as water concentration increases. As ferric-ion concentration increases during acid cleaning of industrial systems, the corrosion rate of steel increases rapidly.

Thus damage locations, although usually widespread, are dependent on many factors. Metallurgy, deposition, design, temperature, pH, water content, dissolved-ion concentration, flow, and other factors all influence attack.

Critical Factors

Intermittent attack

Acid corrosion is often caused by an upset, or it occurs intermittently. Consequently, periods of attack may be separated by intervals of virtually no acid corrosion.

Fresh acid attack is recognized by the absence of corrosion product in wasted areas and the sharpness of attack. Oxide layers are usually easily stripped by a test drop of hydrochloric acid in freshly corroded areas. Deposits are almost always absent. Edges of attacked areas are sharp and angular, as intervening corrosion has not recently occurred. In stainless steels such distinctions blur, as corrosion in intervening periods is usually slight.

Mixed acids

The corrosivity of solutions containing more than one acid may be unrepresentative of the corrosivity of either acid alone. Such mixtures are used widely in a variety of industrial applications.

Corrosion products

Acid attack causes damage directly by wastage and indirectly by increasing and/or moving deposits. For example, during a severe acid excursion, large amounts of iron, copper, zinc, or other materials may be solubilized. Resulting corrosion products will be moved and can possibly foul critical cooling areas such as heat exchanger tubing. Blockage of pipes, screens, and other passages may occur due to solid material sloughed from corroded and fouled surfaces.

Identification

Attack by strong acids tends to produce intense localized wastage. Weak acids cause more general corrosion. However, numerous exceptions to these general rules exist. Fortunately, corrosion damage caused by acids has many unique features that allow easy recognition in most cases.

Strong acids

Pit growth.   When most metals are exposed to highly acidic conditions, pitting results (Fig. 7.2). A pit has a depth greater than its width. In strong mineral-acid corrosion, not only is depth greater than width, but undercutting is pronounced (Fig. 7.3). Cavernous chambers form just below the surface (Fig. 7.4) because of segregation of aggressive anions within the pits.

When corrosion begins, a small amount of positive metal ion is put into solution at the corroding surface. The positively charged ion attracts negatively charged ion from the acid solution. Hydrolysis in the growing pit produces ever-decreasing internal pH, leading to more metal dissolution. The pit begins to undercut, reflecting the increasingly acidic conditions within. The pH within active pits may fall to much lower values than bulk water pH. Because pH remains low, corrosion products do not readily precipitate within the pit. Hence, pits usually contain only small amounts of corrosion product, although

Figure 7.2 Carbon steel pipe attacked by a strong mineral acid.

Figure 7.3   As in Fig. 7.2. Note the narrow pit mouths and the pronounced undercutting.

areas surrounding pits may contain precipitates formed by normal pH-precipitation processes or by oxidation of ions leaking out of pits.

A striking feature of mineral-acid corrosion is the directionality of pit growth. Pits tend to grow in the direction of gravity, even when flow is present. Because pits entrap metal ions and acidic anions, the density of fluid within can be much greater than bulk water density. Pits forming on roof or ceiling surfaces tend to drain by convection that is stimulated by the density difference between the fluid within the pit and the bulk water. This process tends to decrease acidity within pits. Thus, pits growing in the direction of gravity are favored.

Pits form grooves or furrows on wall surfaces on stainless steels (Figs. 7.5 and 7.6). Contents leak out of pit interiors and depassivate areas immediately below the active corrosion sites. Grooves form, fol-

Figure 7.4   Cross section through a mild steel pipe wall suffering severe acid attack.

Figure 7.5 Pits on a large-diameter austenitic stainless steel pipe. Note the grooves formed by leakage of pit contents down the pipe wall in the direction of gravity. (Pipe wall was vertical in service.)

Figure 7.6 As in Fig. 7.5. Note the multiple parallel striations within the larger grooves following stress lines remaining from the original pipe manufacturing processes.

lowing the direction of gravity or flow (also see Figs. 2.24 and 2.25). Elongation of pit mouths may occur because of preferential attack along microstructural irregularities.

Pit morphology varies widely with acid strength, metallurgy, and many other factors. As acids become weaker, pit interiors tend to become less jagged and more rounded. Also, flow tends to smooth pit edges. It is interesting to note that all corrosion begins as localized attack (since some area on the corroding metal surface must be cathodic for corrosion to occur). Thus, in early stages of acid attack it is not at all clear that pits will eventually develop, even when localized attack is present.

Grooving.   Strong acid corrosion will often highlight microstructural features in steel and other formed and rolled alloys. Striations and grooves appear as the acid attacks microstructural irregularities, such as manganese sulfide inclusion "stringers" elongated by rolling or forming processes (Fig. 7.7) or more highly cold-worked regions (Fig. 7.8). Mutual intersection of pits can lead to a jagged surface contour.

Irregular grooving can occur, especially on copper alloys after acid cleaning. Tubes can be only partially filled with cleaning solution. Condensation and running of the fluid down the tube interior cuts tortuous channels (Fig. 7.9).

Figure 7.7 Surface striations and shallow grooves on a carbon steel case coupling. Acid etched the metal along microstructural irregularities producing the grooves. Compare to Fig. 7.8. (Magnification: 7.5×.)

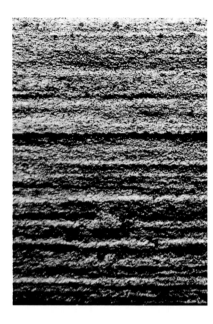

Figure 7.8 Striations on the internal surface of admiralty brass condenser tube after acid cleaning. (Magnification: 15×.)

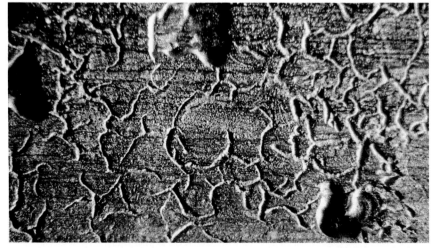

Figure 7.9 Network of shallow ditches caused by condensation of acidic fluids during acid cleaning of a copper condenser tube.

Weld attack.   Welds are often more susceptible to corrosion than other areas (see Chap. 15, "Welds Defects"). Welds may contain porosity, crevices, high residual stresses, and other imperfections that favor attack. Carbon steel welds are usually ditched by acid attack (Fig. 7.10).

Weak acids

Carbonic acid.   Carbonic acid (as do most weak acids) produces smoother attack than stronger acids. Pitting and surface roughness on carbon steel increase as the amount of dissolved oxygen increases (Figs. 7.11 and 7.12). In regions of condensation, grooving frequently occurs (Figs. 7.13 through 7.15). Grooves are caused by the flow of acidic condensate across surfaces. Friable, light-colored corrosion products containing carbonate are often present and sometimes form during cooling. Similar products may surround leaks and failure sites. Carbonate-containing material effervesces strongly when exposed to mineral acids.

Attack is frequently severe when cooling waters contain large amounts of condensate, since carbon dioxide is a condensable gas. Attack is most severe at the point of condensation, before mixing with the cooling water.

Organic acids.   Organic acid strength tends to increase as molecular weight decreases. Hence, low molecular weight organics such as formic acid are quite corrosive relative to longer-chain acids. Additionally, and just as in carbonic acid corrosion, oxygen tends to increase corrosiveness of many organic acid solutions on carbon steels.

Pits (if formed) tend to be shallower, rounder, and less severely undercut than surfaces attacked by strong acids (Fig. 7.16). Wasted

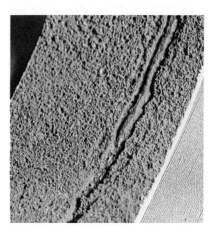

Figure 7.10 Severely attacked weld on a large-diameter steel pipe by an acid upset.

Figure 7.11 Smooth attack on carbon steel by carbonic acid. Note the vivid red color of hematite, indicating high oxygen concentrations.

surfaces are typically rough to jagged, but they become smoother and more undulating as acid strength, oxygen concentration, and potentially corrosive metal-ion concentration decrease.

Flow effects may be pronounced. High-turbulence areas can become preferred attack sites (Fig. 7.17). Erosion-corrosion phenomena are important (Fig. 7.18) (see Chap. 11, "Erosion-Corrosion").

Elimination

Stopping acid corrosion requires the following prevention and emergency action steps:

Prevention

- Monitor water routinely for pH excursions in all large cooling systems.

- Visually inspect acid feed equipment such as pumps on a regular basis.

- Isolate and identify possible sources of in-leakage, such as perforated heat exchangers or makeup water contamination.

- Eliminate windborne acid-gas contamination of cooling towers and other water sources.

Figure 7.12 Jagged fir-tree pattern of metal loss caused by condensing steam with high concentrations of carbon dioxide and oxygen.

- Regularly (at least annually) review all chemical feed and handling equipment.
- Clean systems containing acidic deposits frequently.

Emergency procedures

1. Begin blowdown or drain system if possible.
2. Suspend all acid feed.
3. Monitor pH frequently at appropriate locations. If the pH is less than 3, follow steps 4–8.
4. Block off all acid feed.
5. Begin maximum allowable blowdown immediately.
6. (optional) Add dispersant chelant to remove soluble iron (to prevent oxide hydroxide from forming).
7. (optional) Raise pH by adding appropriate caustic substances.
8. Drain or ensure flow in closed systems or in stagnant areas.

Figure 7.13   Vertical carbon steel gas cooler with grooving at bottom of the last three coils. Carbonic acid-containing condensate attacked.

Figure 7.14   As in Fig. 7.13. Note the presence of orange iron oxides and carbonate. Also note how attack starts as small pits due to dropwise condensation.

Figure 7.15   General thinning along the bottom of a steel conduit carrying carbonic acid-laden water.

Figure 7.16   Carbon steel corroded by organic acid on a cast iron pipe elbow.

Figure 7.17 Locally wasted region at an inlet pipe. See Fig. 7.18. Note the severe attack at the discharge end of the straight pipe.

Figure 7.18 Generally wasted pipe (to the viewer's right) and new pipe. Wastage was general near a pump discharge due to high flow.

Cautions

Oxygen pitting and erosion-corrosion frequently produce damage resembling acid attack. Oxygen pitting can usually be differentiated from acid corrosion by analyzing surface corrosion products and deposits and by carefully examining pit geometry. On nonstainless steels, oxygen pits have less undercut interiors than acid pits and are usually lined with reddish-brown corrosion products. The corrosion products formed by oxygen corrosion are much thicker and more difficult to remove than those found in acid pits.

Pitting on stainless steels is almost always associated with acid conditions that develop within pits by autocatalytic processes. Solution pH outside the pit may be only mildly acidic or not acidic at all (as in the case of sea water, for example).

Cavitation produces spongy, porous-appearing surfaces, strongly resembling acid attack. However, cavitation usually causes highly localized areas of metal loss, unlike acid, which attacks over a much wider area.

Related Problems

Also see Chap. 2, "Crevice Corrosion"; Chap. 5, "Oxygen Corrosion"; Chap. 11, "Erosion-Corrosion"; and Chap. 12, "Cavitation Damage."

CASE HISTORY 7.1

Industry:	Primary metals
Specimen Location:	Valving and bracket
Specimen Orientation:	Horizontal
Environment:	Mill coolant, 140°F (60°C)
Time in Service:	8 years
Sample Specifications:	Carbon steel

Virtually all steel components contacting coolant were attacked. A check valve from a spray system and a hanger bracket were both severely pitted and corroded. The mill components were corroded after an upset that permitted hydrochloric acid to mix with the mill coolant. The pH was at or below 2 during the upset, which continued for at least several weeks.

Valves, nozzles, and regions experiencing high flow were severely attacked (Fig. 7.19). In more quiescent regions, wastage was general with superimposed pitting (Fig. 7.20).

Normal mill coolant pH was near 5. The upset caused large amounts of iron corrosion products to be swept into the coolant. Settling of iron oxides and hydroxides fouled many mill components.

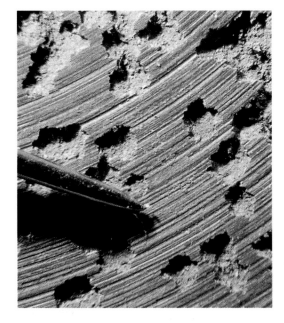

Figure 7.19 Pits on a carbon steel check valve. Note how pits intersect to form areas of jagged metal loss. A steel probe tip is in the photo. (Magnification: 7.5×.)

Figure 7.20   Close-up of general wastage on a carbon steel bracket subjected to low-pH (<2) mill coolant.

CASE HISTORY 7.2

Industry:	Steel
Specimen Location:	Outlet piping of a sulfuric acid storage tank
Specimen Orientation:	Horizontal
Environment:	Internal: Concentrated sulfuric acid, 70–100°F (21–38°C)
Time in Service:	~1 year
Sample Specifications:	½ in. (1.3 cm) outer diameter, austenitic stainless steel tubing, 0.030 in. (0.076 cm) wall thickness

Feed lines from a concentrated sulfuric acid storage tank failed. Tubing contained large, ragged holes due to severe internal surface wastage (Fig. 7.21).

The internal surface was generally covered with dark, tenacious corrosion products. Traces of greenish-white corrosion product were also present. Many small, intersecting pits pockmarked internal surfaces. Thinning was pronounced near the failure where the remaining metal was paper thin.

The failure was caused by sulfuric acid containing excessive water. Nearly pure sulfuric acid is only very mildly corrosive. However, when moisture content approaches 20% at 100°F (38°C), corrosion rates on 304 stainless steel are 1 in. (2.5 cm) per year. At 30% moisture content, corrosion rates are an order of magnitude greater.

It was likely that water entered the storage tank due to condensation of moisture-laden air that was sucked into the tank. As night fell and temperature decreased, moisture would condense and dilute the acid. Air-drying systems were not in good operating order. Suggestions were made to test acid water content, and if the acid was found to contain 10% water or more, acid was to be added to reduce water concentrations to below 5%.

Figure 7.21   Austenitic stainless steel pipe that carried sulfuric acid. Failure was caused by severe general wastage from internal surfaces.

CASE HISTORY 7.3

Industry:	Steel
Specimen Location:	Electrostatic precipitator
Specimen Orientation:	Vertical
Environment:	Electrostatic precipitator receives air from steel blast furnace; 50–180°F (10–82°C); high moisture, chlorides, sulfides and sulfate, iron oxides
Time in Service:	6 months
Sample Specifications:	¾ in. by 2 in. (1.9 cm by 5.1 cm), 304 stainless steel corrosion coupon

Austenitic stainless steel coupons were placed in a large electrostatic precipitator. Each coupon rapidly developed pits. Attack was caused by chlorides dissolved in acidic aqueous solutions.

Small pits, some the size of pinheads, pockmarked surfaces (Fig. 7.22). Each pit was surrounded by a halo of reddish-brown rust. Cyclic wetting and evaporation caused chloride concentration and increased acidity locally.

Coupon tests involved a number of metallurgies and were done to evaluate precipitator-plate alloys. Test stainless steel plates failed, not only because of pitting but also because stress-corrosion cracks developed.

Figure 7.22   Pitting on a 304 stainless steel coupon caused by acidic, chloride-containing water condensating and evaporating. High chloride concentrations were produced locally.

CASE HISTORY 7.4

Industry:	Aluminum
Specimen Location:	Coolant base connector
Specimen Orientation:	Unknown
Environment:	Rolling oil, mill coolant
Time in Service:	8–10 weeks
Sample Specifications:	¾ in. (1.39 cm) inner diameter, free-machining carbon steel

Severe corrosion by turbulent mill coolant was found generally throughout a rolling-oil system. Hose couplings were severely wasted in as little as 8 weeks (Fig. 7.23*A* and *B*). Turbulence caused by high-velocity flow through nozzles accelerated attack. Attack at bends, elbows, intrusive welds, and discharge areas was also severe.

Surfaces were covered with shallow, pitlike depressions intersecting to cause general wastage. Wall thickness was reduced by at least 75% in places. Threads were severely attacked. Corrosion products were absent, and only bare metal was present in attacked regions.

Wastage was caused by exposure to oleic acid and short-chain organic acids in the rolling oil. Fatty acids break down to form shorter-chain acids in service. However, oleic acid, of and by itself, is fairly corrosive. Attack due to oleic acid can be reduced substantially using appropriate chemical inhibition.

Figure 7.23A   Free-machining carbon steel hose coupling after 2 months in a solution containing several percent oleic acid.

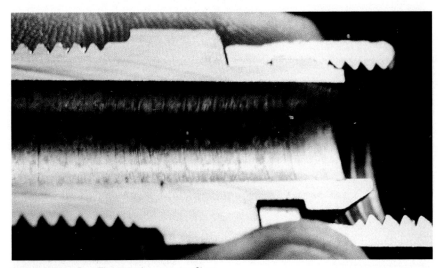

Figure 7.23B   See Fig. 7.23A; new coupling.

CASE HISTORY 7.5

Industry:	Pulp and paper
Specimen Location:	Compressor aftercooler tube
Specimen Orientation:	Horizontal
Environment:	Internal: 100°F (38°C) max., unregulated flow, 0.01 ppm ammonia in water
	External: Air
Time in Service:	3 years
Sample Specifications:	⅝ in. (1.6 cm) outer diameter, copper tubes

Visual examination of external surfaces revealed grooves and general metal loss (Figs. 7.24 and 7.25). Metal loss was caused by erosion-corrosion. However, the corrosive loss was more important than erosive loss, since metal loss was also substantial in low-flow regions.

Corrosion was caused by carbonic acid. A film of condensed moisture and dissolved carbon dioxide formed the acid. The erosion was caused by high-velocity movement of air across the tubes. Attack occurred intermittently. Deepest metal loss was 33% of the 0.040 in. (0.10 cm) wall thickness.

Figure 7.24 Deep groove in region of localized metal loss on a copper tube. Grooving was caused by carbonic acid containing high concentrations of dissolved oxygen.

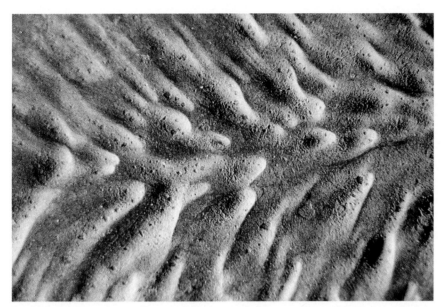

Figure 7.25 Smooth, wavelike contour in a region of general metal loss on copper due to carbonic acid. The groove patterns indicate channeling of corrosive fluids by flow. (Light is coming from the reader's right.)

Alkaline Corrosion

General Description

Corrosion of industrial alloys in alkaline waters is not as common or as severe as attack associated with acidic conditions. Caustic solutions produce little corrosion on steel, stainless steel, cast iron, nickel, and nickel alloys under most cooling water conditions. Ammonia produces wastage and cracking mainly on copper and copper alloys. Most other alloys are not attacked at cooling water temperatures. This is at least in part explained by inherent alloy corrosion behavior and the interaction of specific ions on the metal surface. Further, many dissolved minerals have normal pH solubility and thus deposit at faster rates when pH increases. Precipitated minerals such as phosphates, carbonates, and silicates, for example, tend to reduce corrosion on many alloys.

Certain alloys frequently used in cooling water environments, notably aluminum and zinc, can be attacked vigorously at high pH. These metals are also significantly corroded at low pH and thus are said to be *amphoteric*. A plot of the corrosion behavior of aluminum as a function of pH when exposed to various compounds is shown in Fig. 8.1. The influence of various ions is often more important than solution pH in determining corrosion on aluminum.

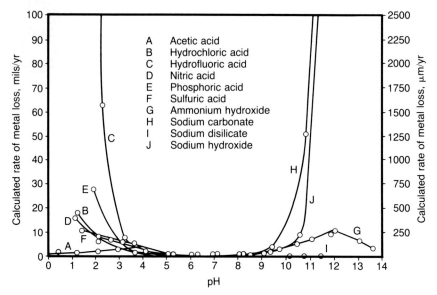

Figure 8.1 Effect of pH on corrosion of 1100-H14 alloy (aluminum) by various chemical solutions. Observe the minimal corrosion in the pH range of 4–9. The low corrosion rates in acetic acid, nitric acid, and ammonium hydroxide demonstrate that the nature of the individual ions in solution is more important than the degree of acidity or alkalinity. *(Courtesy of Alcoa Laboratories; from* Aluminum Properties and Physical Metallurgy, *ed. John E. Hatch, American Society for Metals, Metals Park, Ohio, 1984, Figure 19, page 295.)*

Aluminum

Thermodynamically, aluminum should be a highly reactive metal. However, reactivity is limited in most natural environments. When exposed to water or water and air, aluminum quickly forms a protective oxide layer. Once formed, the oxide slows further corrosion. This oxide layer may be as thin as about 5×10^{-9} m (50 Å) when formed naturally in air, but it is thicker when formed in water and can be made up to about 3000 times thicker by anodizing.

Aluminum corrodes in water as in Reaction 8.1:

$$2Al + 6H_2O \rightarrow Al_2O_3 \cdot 3H_2O + 3H_2 \uparrow \qquad (8.1)$$

The final corrosion product, aluminum oxide trihydrate, is called Bayerite.

At a pH above about 9, in the presence of sodium carbonate or sodium hydroxide, for example, the protective oxide layer is rapidly dissolved and corrosion becomes severe (Fig. 8.1). Aluminum in the presence of sodium hydroxide corrodes as in Reaction 8.2:

$$Al + NaOH + H_2O \rightarrow NaAlO_2 + 3/2H_2 \uparrow \qquad (8.2)$$

Note: At high temperatures, concentrated caustic can corrode steel, producing $NaFeO_2$ and Na_2FeO_2.

Figure 8.1 shows that aluminum is corroded by alkaline substances, albeit at different rates, when water pH exceeds 9. Corrosion by inorganic salts between a pH of 5 and 9 is very slow near room temperature. Aluminum shows no significant corrosion in most natural waters up to about 350°F (180°C). Of course, natural waters vary widely in composition, and exceptions do occur.

Whatever the water composition, corrosivity can be increased by evaporation, which may elevate pH by increasing concentrations of ions in the remaining liquid. This is the reason that cooling systems that experience boiling and/or large evaporative losses without sufficient makeup water additions may be especially prone to attack. Cycling may also increase dissolved species concentrations.

Zinc

Zinc is attacked at high pH. However, in weakly alkaline solutions near room temperature, corrosion is actually very slight, being less than 1 mil/y (0.0254 mm/y) at a pH of 12. The corrosion rate increases rapidly at higher pH, approaching 70 mil/y (1.8 mm/y) at a pH near 14. Just as in aluminum corrosion, protection is due primarily to a stable oxide film that forms spontaneously on exposure to water. High alkalinity dissolves the oxide film, leading to rapid attack.

Copper alloys

Copper alloys are attacked at high pH. However, attack is usually caused not by elevated pH alone but because of copper complexation by ammonia or substituted ammonium compounds. In fact, copper resists corrosion in caustic solutions. For example, corrosion rates in hot caustic soda may be less than 1 mil/y (0.025 mm/y).

Copper-alloy corrosion behavior depends on the alloying elements added. Alloying copper with zinc increases corrosion rates in caustic solutions whereas nickel additions decrease corrosion rates. Silicon bronzes containing between 95% and 98% copper have corrosion rates as low as 2 mil/y (0.051 mm/y) at 140°F (60°C) in 30% caustic solutions. Figure 8.2 shows the corrosion rate in a 50% caustic soda evaporator as a function of nickel content. As is obvious, the corrosion rate falls to even lower values as nickel concentration increases. Caustic solutions attack zinc brasses at rates of 2 to 20 mil/y (0.051 to 0.51 mm/y).

Ammonium hydroxide and substituted ammonium compounds corrode copper alloys. Wastage is faster when oxygen concentration and/or

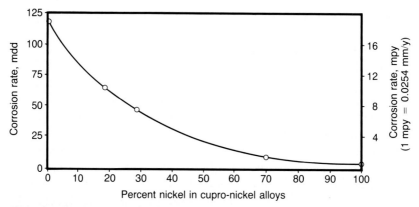

Figure 8.2   Corrosion rates of copper and copper-nickel alloys in a 50% caustic soda evaporator *(Courtesy of F. L. LaQue, Corrosion 10:391:1954.)*

temperature increase. Corrosion rates of several copper alloys in a 0.8% ammonia solution are given in Table 8.1.

Nickel additions reduce corrosion due to both caustic solutions and ammonia. Monel is resistant to attack in ammonia-containing waters and vapors. Reactions with ammonia are as follows:

Anode:
$$Cu + 4NH_3 \rightarrow Cu(NH_3)_4^{++} + 2e^- \tag{8.3}$$

Cathode:
$$Cu(NH_3)_4^{++} + 2e^- \rightarrow Cu(NH_3)_2 + 2NH_3 \tag{8.4}$$

or
$$2Cu(NH_3)_4^{++} + H_2O + 2e^- \rightarrow Cu_2O + 2NH_4^+ + 6NH_3 \tag{8.5}$$

Zinc brasses are corroded much more rapidly by ammonium hydroxide than by caustic solutions. Corrosion rates approaching 240 mil/y (6.1 mm/y) have been measured at room temperature in two normal ammonium-hydroxide solutions. Corrosion rates in hot, concentrated caustic solutions may be as high as 70 mil/y (1.8 mm/y).

TABLE 8.1   Corrosion Rates of Several Copper Alloys in 0.8% Ammonia at 104°F (40°C)

Alloy	Corrosion rate		
	MDD	MPY	MM/Y
Copper	85	14	0.36
Cartridge brass (70:30 Cu-Zn) 260	43	7	0.2
Gun metal (88:10:2 Cu-Sn-Zn) 905	30	5	0.1
Copper-manganese alloy (95:5 Cu-Mn)	9	2	0.05

SOURCE: After J. A. Radley, J. S. Stanley, and G. E. Moss, *Corrosion Technol.* 6:229:1959.

Locations

Wastage is pronounced in equipment contacting high-pH fluids. Chemical process equipment, heat exchangers, water-cooled process reactors, valving, transfer pipes, and heating and cooling systems are often affected.

Concentrating effects associated with the condensation of aggressive gases such as ammonia frequently cause elevated pH. Condensers, especially those tubed with copper alloys, suffer wastage. Evaporation can concentrate caustic species, and attack may occur. Also, concentration at crevices and beneath deposits can accelerate attack. Aluminum and zinc alloys (e.g., galvanized sheet metal) used in automotive cooling components, cooling towers, electrical systems, sacrificial anodes, and valve components are often attacked. Steel components are almost never attacked in alkaline cooling water systems unless temperatures become high and caustic concentrations are very large.

Critical Factors

Aluminum alloys are corroded at both high and low pH. Not all compounds that increase pH cause severe attack. Ammonium hydroxide only moderately increases corrosion rates. Wastage actually decreases above a pH of 12 in ammonium hydroxide solutions (see Fig. 8.1). A caustic solution causes corrosion rates to increase substantially as pH rises. The Al^{+3} ion reacts vigorously with OH^- to produce AlO_2^-.

Copper alloys are attacked by ammonium hydroxide, but corrosion rates increase only moderately with caustic concentration.

Just as at low pH, concentration mechanisms substantially increase attack. The two principal mechanisms of concentration are evaporation and condensation. Evaporation increases solute concentrations of compounds with vapor pressures lower than water (such as caustic compounds). Condensation increases concentration of aggressive gases such as ammonia.

Identification

Aluminum alloys

Wastage morphology is vague. Usually broad areas contacting high-pH fluids are affected (Fig. 8.3). Well-defined pits are rare. Intersecting areas of shallow metal loss produce generally wasted surfaces. More localized attack is found in early stages of corrosion and on castings (Figs. 8.4 and 8.5). Wastage progresses so that individual areas of attack eventually merge.

Figure 8.3   General wastage of an aluminum water manifold from a diesel engine cooling system. Note the generally wasted internal surface due to concentrated caustic and the presence of white deposits and corrosion products.

Figure 8.4   Corroded aluminum-alloy casting valve seat. The seat was in contact with water having a pH between 9 and 10 for 3 months. Red marks were drawn on the surfaces to assist metallographic preparation.

Figure 8.5 As in Fig. 8.4 but at greater enlargement. Note the intersection of pitlike depressions to form more general metal loss.

White, friable corrosion products composed of Bayerite $Al_2O_3 \cdot 3H_2O$, caustic, and $NaAlO_2$ cover corroded areas (Fig. 8.3). The white corrosion product and deposit usually test as distinctly alkaline when mixed with distilled water. Corrosion products usually cling tenaciously to the underlying metal and do not form voluminous lumps. Instead, corrosion products line and coat generally wasted surfaces below.

Copper alloys

Wastage may appear as general attack or as grooving, or it may be localized. Pitting almost never occurs. Each attack morphology is characteristic of concentration processes associated with corrosion.

In condensers with steam on external tube surfaces, grooving is common on shell sides. Grooving is usually confined to regions near tube sheets and baffles. Condensed vapors containing high concentrations of ammonia collect on and run down the tube sheets and baffles, collecting where tubes are rolled into the sheets or pass through baffles. Broad annular grooves are cut into surfaces abutting the sheets and baffles (Fig. 8.6). Frequently, sinuous grooves run in the direction of gravity (Fig. 8.7). Grooves are produced by rivulets of aggressive condensate and may superficially resemble stress-corrosion cracks (see Chap. 9, "Stress-Corrosion Cracking"). Inside piping where condensation occurs, grooving can be pronounced (Fig. 8.8). In extreme cases, attack can become general, producing undulating surface contours (Figs. 8.9*A* and *B*).

Figure 8.6 Severe grooving by ammonia-containing condensate on an admiralty brass condenser tube. Condensate flowed down the tube sheet and collected at the tube end, cutting an annular groove. Note the perforation just outside the tube roll.

Figure 8.7 Small grooves on the tube shown in Fig. 8.6. Grooves are condensate rivulet paths cut into the tube surface. (Magnification: 10×.)

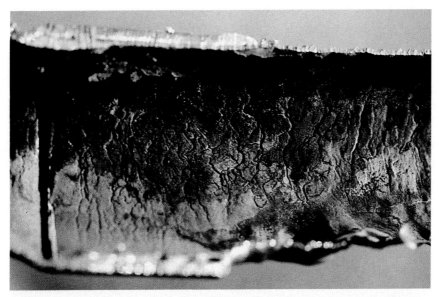

Figure 8.8 Severe grooving on the internal surface of a copper pipe carrying condensate. Grooves were cut by condensing vapors running down pipe walls. Note the vivid blue corrosion products and deposits near the bottom.

Grooving becomes broad along the bottom of horizontal piping where ammonia-rich condensate collects. One often finds a zone of severe general thinning along the bottom of pipes carrying condensate (Fig. 8.10). The zone of smooth metal loss will occur on the bottom of pipes and will be bordered on either side by sharp lines marking the steam-water interface. Frequently there is grooving in the direction of gravity immediately above the generally thinned zone. Grooves are formed as rivulets of condensed vapor, laden with ammonia, run down vertical surfaces (Figs. 8.6 through 8.8).

Other than grooving, localized attack due to ammonia is relatively rare. Patches of attack can occur in conjunction with biological fouling and decomposition of organic materials, which generate ammonia.

Voluminous corrosion products are usually absent, as most copper amine complexes are quite soluble. Adjacent to corroded areas, one often finds small amounts of corrosion products and deposits colored a vivid blue-green by compounds containing liberated copper ion.

Elimination

Decreasing attack can involve pH control and chemical inhibition. Sodium silicate markedly improves the corrosion resistance of aluminum

Figure 8.9*A*   A brass condenser tube severely wasted by condensing vapors containing ammonia.

Figure 8.9*B*   As in Fig. 8.9A. Note the fine grooves resembling cracks. Compare to Figs. 8.6 through 8.8.

Figure 8.10   A large, ragged perforation in a region of severe internal wastage on a copper pipe. Corrosion was caused by ammonia-containing condensate.

alloys in many alkaline solutions. Filming amines may have some beneficial effect on yellow-metal corrosion in ammonia solutions.

Cautions

Reducing pH usually has a beneficial effect on corrosion caused by alkaline substances. However, this seemingly obvious solution has a number of drawbacks. Chemical treatment programs work most effectively in certain pH ranges. Decidedly acidic waters can cause corrosion problems as bad or worse, albeit different, than those caused by alkaline waters. Finally, if concentration mechanisms such as evaporation or condensation are present, merely decreasing pH may prove ineffective in controlling attack.

Related Problems

See also Chap. 2, "Crevice Corrosion"; Chap. 4, "Underdeposit Corrosion"; and Chap. 7, "Acid Corrosion."

CASE HISTORY 8.1

Industry:	Polyethylene film manufacture
Specimen Location:	Cooling water conduit
Specimen Orientation:	Vertical
Environment:	Cooling water, 70–100°F (21–38°C)
Time in Service:	Less than 1 year
Sample Specifications:	1 in. (2.5 cm) diameter, aluminum alloy tubing, anodized

An aluminum alloy cooling water conduit section was removed from a machine that made polyethylene sheeting. A large hole had formed just below a ¾ in. (1.9 cm) cooling water orifice. The designed orifice is shown in Fig. 8.11, with the elliptically shaped corroded hole just below.

The designed orifice allowed water to exit and move down an annular space between the conduit and a sheathing pipe. Impingement and the relatively high water pH (9.5–10.5) caused corrosion of the designed orifice edges and localized wastage, producing the elliptical perforation.

The accelerating effect of turbulence in alkaline attack on aluminum alloys is well illustrated in this case history.

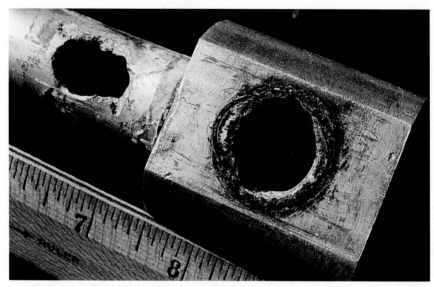

Figure 8.11   An aluminum cooling water conduit severely attacked by a caustic in service. The large circular hole in the machined face is corroded around its edges. The elliptical hole between the 6- and 7-in. ruler markings was formed by corrosion penetrating the conduit wall.

CASE HISTORY 8.2

Industry:	Metals fabrication
Specimen Location:	Annealing machine sleeve
Specimen Orientation:	Vertical
Environment:	Internal: Cooling water containing neutralizing amine, oxygen scavenger, caustic
Time in Service:	1 month
Sample Specifications:	3 in. (7.6 cm) aluminum tubing

Aluminum sleeves used in an annealing machine to heat treat copper rod failed rapidly (Figs. 8.12 and 8.13). Attack was caused by a caustic that concentrated due to evaporation of water in the annealing sleeve.

Water taken from below these sleeves had a pH of 10.2. It was estimated that water pH must have been near 14 at the metal surface.

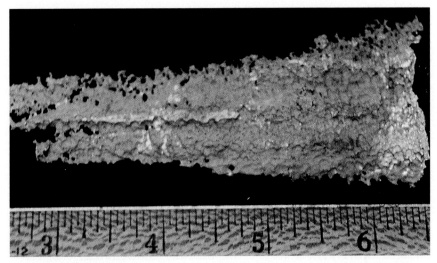

Figure 8.12 A fragment of a severely corroded aluminum annealing sleeve. Attack was by caustic, which caked the specimen.

Figure 8.13 As in Fig. 8.12, but a threaded portion of the annealing sleeve. Note the heavy caustic deposits.

CASE HISTORY 8.3

Industry:	Utility
Specimen Location:	Low-pressure turbine condenser inlet just outside tubesheet
Specimen Orientation:	Horizontal
Environment:	Internal: pH 8.2–7.8, phosphate-zinc, chlorine, 65–100°F (18–38°C), calcium, hardness 300–600 ppm, total alkalinity 45–60 ppm
	External: (a) Saturated steam 105–130°F (41–54°C), NH_3 < 0.3 ppm, pH 8.7–9.0, dissolved oxygen 10–20 ppb
	(b) (Startup, once or twice a year), NH_3 up to 75 ppm, pH 8.7–9.4, dissolved oxygen up to 3500 ppb, hydrazine 10–20 ppb
Time in Service:	14 years
Sample Specifications:	1 in. (2.5 cm) outer diameter, admiralty brass tubing

The tube section was one of several that contained a perforation in a region of severe external surface metal loss (Fig. 8.6). Wall thickness was as great as 0.040 in. (0.10 cm) on the rolled end. Just outside the rolled end, wall thickness dropped to a maximum of 0.030 in. (0.076 cm). Most areas were as thin as 0.016 in. (0.042 cm).

Close visual inspection of the thinned surface revealed small, longitudinally oriented grooves and gullies running from the rolled tube end across the thinned region (Fig. 8.7). The grooves were deepest at the rolled tube end and branched as they moved away from the tube sheet. The external surface was covered with a thin, dark oxide layer. The layer was replaced in places by a dark green patina. In the thinned region, the surface had an undulating contour.

Wastage was caused by waters containing high concentrations of ammonia. Steam containing high levels of ammonia condensed, ran down the tube sheet, and collected at tube ends, causing pronounced localized wastage.

Cracking Phenomena
in Cooling Water Systems

Of the numerous failures that can plague cooling water systems, none is more insidious than cracking. Cracking often occurs with little or no warning and may require an immediate shutdown of equipment. Generally, it is a very subtle form of metal damage, typically involving little, if any, metal loss.

A crack may be defined as a continuous separation in a metal component. The conditions under which cracking occurs are many and varied. However, one condition is necessary, although not necessarily sufficient, for all cracking mechanisms—stress. Stress may be residual and/or applied, static and/or cyclic, and of a high or low level.

Most cracking problems in cooling water systems result from one of two distinct cracking mechanisms: stress-corrosion cracking (SCC) or corrosion fatigue.

SCC and corrosion fatigue can be described as environmental cracking phenomena because they require, in addition to a tensile stress, a corrosion reaction. If either of these components is eliminated, cracking of this type will not occur. Cracks of this type generally produce brittle (thick-walled) fractures even in ductile metals. The cracks are typically tight and difficult to detect with the unaided eye. Cracks may occur alone or in groups and may be branched or unbranched depending on the cracking mechanism. Branching is very often present in SCC, and lack of branching is typical of corrosion fatigue.

SCC and corrosion fatigue differ in the nature of the stresses and corrodents that cause them. SCC occurs under static and/or cyclic tensile stresses. Corrosion fatigue requires a cyclic stress, although a static stress may contribute. SCC requires exposure to a specific corrodent; corrodents for corrosion fatigue are nonspecific.

Stress-Corrosion Cracking

General Description

Few, if any, failure mechanisms have received as much attention as stress-corrosion cracking (SCC). Yet despite an enormous research effort over many years, an acceptable, generalized theory that satisfactorily explains all elements of the phenomenon has not been produced. SCC is a complex failure mechanism. Nevertheless, its basic characteristics are well known, and a wealth of practical experience permits at least a moderately comfortable working knowledge of the phenomenon.

SCC has been defined as failure by cracking under the combined action of corrosion and stress (Fig. 9.1). The stress and corrosion components interact synergistically to produce cracks, which initiate on the surface exposed to the corrodent and propagate in response to the stress state. They may run in any direction but are always perpendicular to the principal stress. Longitudinal or transverse crack orientations in tubes are common (Figs. 9.2 and 9.3). Occasionally, both longitudinal and transverse cracks are present on the same tube (Fig. 9.4). Less frequently, SCC is a secondary result of another primary corrosion mode. In such cases, the cracking, rather than the primary corrosion, may be the actual cause of failure (Fig. 9.5).

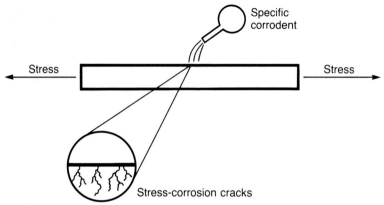

Figure 9.1   Stress-corrosion cracks result from the synergistic interaction of tensile stress and a specific corrodent.

The surface from which the cracks originate may not be apparent without a microstructural examination. Stress-corrosion cracks invariably produce brittle (thick-walled) fractures regardless of the ductility of the metal.

SCC typically produces tight, branched cracks that may be difficult to detect with the unaided eye. Staining or accumulation of corrosion products may or may not accompany cracking.

It is important to realize that the conditions causing SCC may not only occur during normal operation of equipment but also during start-

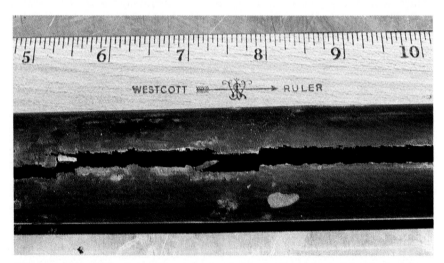

Figure 9.2   Longitudinal stress-corrosion cracks in a heat exchanger tube; the broad gap between the crack faces reveals that high-level residual hoop (circumferential) stresses from the tube-forming operation provided the stress component required for SCC.

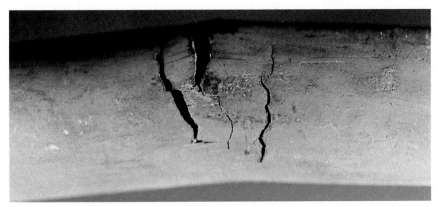

Figure 9.3　Transverse stress-corrosion cracks in a condenser tube; the presence of the cracks along just one side indicates that a bending moment provided the stress.

up, shutdown, idle periods, or system upsets. Stresses and environmental conditions under these circumstances can be quite different than those encountered during normal operation.

Locations

Due to the wide variety of environments to which cooling water components are exposed on the cooling water and process sides, it is difficult to specify favored locations for SCC. However, a few general observations may be permitted:

- With the possible exception of systems using sea water, estuarial waters, and/or industrially contaminated waters for cooling, condi-

Figure 9.4　Both longitudinal and transverse stress-corrosion cracks on a brass heat exchanger tube that had been exposed to ammonia. Note the branching of the cracks.

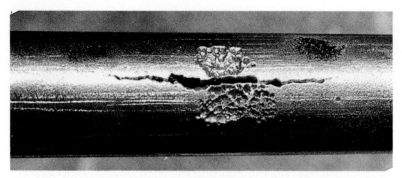

Figure 9.5   Stress-corrosion cracks associated with simple corrosion.

tions for SCC occur more frequently on the process side, especially when copper-based alloys are used for condensing steam. Conditions on the process side frequently provide opportunities for the concentration of specific corrodents.

■ Sites possessing high-residual stresses—such as welded assemblies, rolled-in tube ends—may be susceptible.

■ Sites subject to high-service stresses—such as points of physical constraint—may be susceptible.

■ Sites subject to boiling or evaporation are susceptible due to the potential for concentrating SCC agents.

Critical Factors

The two basic requirements for SCC are as follows:

1. *Sufficient tensile stress.*   Sufficiency here is difficult to define since it depends on a number of factors such as alloy composition, concentration of corrodent, and temperature. In some cases, stresses near the yield strength of the metal are necessary. In other cases, the stresses can be much lower. However, for each combination of environment and alloy system, there appears to be a threshold stress below which SCC will not occur. Threshold stresses can vary from 10 to 70% of yield strength depending on the alloy and environment combination and temperature (Fig. 9.6).

Stresses may be applied and/or residual. Examples of applied stresses are those arising from thermal expansion and contraction, pressure, and service loads. Applied stresses may be continuous or intermittent. Examples of residual stresses are those arising from welding, fabrication, and installation. The importance of residual stresses in SCC should not be underestimated. Residual stresses may

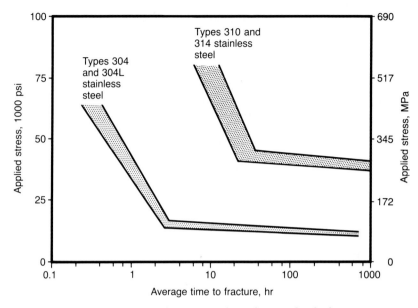

Figure 9.6 Threshold stresses of two types of stainless steel in boiling 42% magnesium chloride solution. (*Reprinted with permission of American Society for Metals from* Metals Handbook, *vol. 10, 8th ed., Metals Park, Ohio, 1974, p. 210.*)

approach the magnitude of the yield stress. Design stresses are typically well below the yield stress.

In general, the propensity and severity of cracking increases as stress levels rise. However, SCC can and does occur within the range of typical service stresses.

2. *A specific corrodent.* One of the unusual and interesting features of SCC is the specificity of the corrodent. A particular alloy system is susceptible to SCC only when exposed to certain corrodents, some or all of which may be unique to that particular alloy system. For example, austenitic stainless steels (300 series) are susceptible to cracking in chloride solutions but are unaffected by ammonia. Brasses, on the other hand, will crack in ammonia but remain unaffected by chlorides. The corrodent need not be present at high concentrations. Cracking has occurred at corrodent levels measured in parts per million (ppm).

Essentially all industrial metals are susceptible to SCC in some specific environment. Of the metals commonly used in cooling water systems, copper-based alloys and stainless steels are most frequently affected. Common specific corrodents causing SCC in these and other heat exchanger metals are listed in Table 9.1.

TABLE 9.1   Stress-Corrosion Cracking Agents

Metal corrodent	
Austenitic stainless steel	Chlorides Hot concentrated caustic Hydrogen sulfide
Carbon steel	Concentrated caustic Concentrated nitrate solutions Anhydrous ammonia Carbonate and bicarbonate solutions
Copper-based alloys	Ammonia (vapors and solutions) Amines Sulfur dioxide Nitrates, nitrites
Titanium	Ethanol Methanol Sea water (certain alloys) Hydrochloric acid (10% solution)

Note that the environments that produce SCC are not necessarily corrosive to the metals in the unstressed state. Other factors that may also influence susceptibility to SCC are listed here; discussion of these items is, however, beyond the scope of this book.

Temperature

pH

Time

Concentration of corrodent

Presence of oxidizing agents

Metal composition and structure

Alternate wetting and drying

Concentration effects

An important but frequently overlooked condition that can result in SCC involves concentration effects. These effects are of two types—stress concentration and corrodent concentration.

Stress concentration.   Stress concentration refers to physical discontinuities in a metal surface, which effectively increase the nominal stress at the discontinuity (Fig. 9.7). Stress-concentrating discontinuities can arise from three sources:

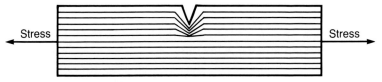

Figure 9.7   Concentration of stress at the root of a notch.

1. Design: Notches, holes, keyways, welds, and so on
2. Fabrication and installation: Mechanical damage, arc strikes, weld-related cracking, rough grinding or machining, and so on
3. Localized deterioration: Corrosion (especially pitting and intergranular attack), erosion, cavitation, mechanical wear, and so on (see Case History 9.8).

Discontinuities intensify stresses. Since susceptibility to SCC commonly increases as stress level increases, stress-corrosion cracks may occur at a discontinuity, whereas smooth areas remain intact.

Corrodent concentration.   Numerous environmental circumstances can lead to the concentration of a corrodent. These can be categorized as follows:

- *Localized stagnation.*   Permeable deposits, crevices, preexisting cracks, and other conditions that result in physical shielding can lead to concentration of a corrodent in the stagnant solution, which can be 10–100 times or more greater than that measured in a bulk fluid (see Case History 9.1).

- *Evaporation.*   Evaporative concentration can produce concentrations of 100,000 times or more in certain circumstances. Heat transfer surfaces, liquid and vapor interfaces, and regions where wetting and drying conditions occur are areas subject to evaporative concentration (see Case Histories 9.1, 9.4, and 9.6).

- *Vapor spaces.*   In addition to the possibility of wetting and drying conditions, vapor spaces may allow gaseous corrodents, such as ammonia and sulfur dioxide, to concentrate to high levels in thin films of condensate (see Case Histories 9.2, 9.3, and 9.8).

The difficulty in accurately estimating the degree of local concentration remains one of the principal reasons susceptibility to SCC in a specific environment or circumstance is difficult to predict. Measurement of nominal stresses or levels of corrodent in the bulk environment can be quite misleading as predictors of SCC susceptibility.

Identification

Visual identification prior to failure is difficult due to the typical tightness of stress-corrosion cracks. A low-power hand lens will greatly aid determination. Crack enhancement may be achieved through the use of dye penetrants. Severe cracking may be detectable using ultrasonic, radiographic, or acoustic emission techniques.

Stress-corrosion cracks tend to branch along the metal surfaces. Typically, evidence of corrosion, such as accumulations of corrosion products, is not observed, although stains in the cracked region may be apparent. Stress-corrosion cracks tend to originate at physical discontinuities, such as pits, notches, and corners. Areas that may possess high-residual stresses, such as welds or arc strikes, are also susceptible.

Elimination

The two principal factors governing SCC are tensile stresses and exposure to a specific corrodent. These two factors interact synergistically to produce cracking. Only one factor needs to be removed or sufficiently diminished to prevent cracking.

Tensile stress

Residual.   Stress-relief-anneal components and assemblies following cold working or welding operations. Note, however, that annealing has no effect on applied stresses.

Applied.   Generally, adequate reduction of applied stresses involves equipment redesign.

Corrodent

If the responsible specific corrodent cannot be entirely removed from the environment, it may be beneficial to reduce its concentration, since susceptibility to SCC is frequently concentration dependent. However, if this method of elimination is chosen, care must be taken to avoid conditions that will increase concentration locally, such as evaporation, localized boiling, alternate wet and dry conditions, crevices, and deposits.

If the corrodent cannot be removed or reduced, the addition of an appropriate inhibitor may prevent cracking. However, complete, effective coverage by an inhibitor may not be achieved or economically feasible. Application of cathodic protection can stifle SCC in some metal systems. (*Precautionary note:* If cracking is associated with hydrogen embrittlement rather than SCC, the use of cathodic protection can intensify cracking.)

Since metals are sensitive to specific SCC agents, successful elimination of SCC problems can be effected by changing the metallurgy of the affected component to a material that will not crack in the existing environment. This method takes advantage of the specificity of the environment and alloy system relationship. However, it is important to identify the specific corrodent responsible for cracking since some corrodents affect more than one alloy system.

Since SCC is often dependent on environmental factors other than stress and exposure to a specific corrodent, appropriate alteration of these other factors may be effective. For example, a reduction in metal temperature, a change in pH, or a reduction in the levels of oxygen or oxidizing ions may reduce or eliminate SCC.

Cautions

Stress-corrosion cracks tend to be fine, tight, and easily overlooked. Various nondestructive techniques are available to aid in the discovery of cracks, such as dye penetrant, and ultrasonic and radiographic techniques.

It is possible to confuse SCC with other brittle cracking phenomena. Confirmation of SCC typically requires a metallographic examination. On thin-walled components, the surface from which the cracking originates may not be apparent. In these cases, a formal metallographic examination may be required to assure positive identification of the surface from which the cracks originate.

Stress-relief-annealing cannot be expected to eliminate SCC in every case. Only residual stresses are reduced in stress-relief-annealing. Applied stresses, which may be responsible for the cracking, will remain. Inhibitors are not 100% effective in combating SCC. Complete coverage and inhibition is difficult to achieve, especially below deposits, in crevices, and in pits.

Due to the complexity of the SCC phenomenon, preventive techniques have been expressed in general rather than specific terms. It should be recognized that exceptions to these general guidelines may exist and that cracking could result in some specific circumstance even if the guidelines are followed.

Cathodic protection can stifle SCC in some metal systems. However, if cracking is the result of hydrogen embrittlement rather than SCC, the use of cathodic protection can intensify cracking.

Related Problems

See Chap. 10, "Corrosion Fatigue."

CASE HISTORY 9.1

Industry:	Chemical process
Specimen Location:	Heat exchanger tube
Specimen Orientation:	Horizontal
Environment:	Internal: A mixture of steam, condensate, hydrogen, nitrogen, and carbon dioxide; outlet temperature 165°F (75°C), inlet temperature 265°F (130°C), condensate pH 10
	External: Treated cooling water, outlet temperature 105°F (40°C), inlet temperature 85°F (30°C), pH 7.5, chloride 150–200 ppm, phosphate 10–12 ppm, zinc 2–4 ppm
Time in Service:	15 years
Sample Specifications:	1 in. (25.4 mm) outer diameter, 304 stainless steel

The transverse (circumferential) fissures illustrated in Figs. 9.8 and 9.9 are similar to failures that had occurred in many tubes. All failures occurred near the tube sheet at the hot end of the exchanger (Fig. 9.10). The initial failure had occurred 3 years earlier.

Figure 9.9 shows the proximity of the cracks to the rolled end of the tube. The cracks occurred from just one side. The original crack was tight, but distortion of the tube during its removal from the exchanger expanded the cracks, giving them a fissurelike appearance in the photographs.

Microstructural examinations revealed that the cracks originated on the external surface. The cracks were branched and ran across the metal grains (transgranular).

The transverse orientation of the cracks along just one side of the tube reveals that bending provided the stresses for cracking. The tube sheet acted as a constraint to the bending, intensifying stresses in the tube wall adjacent to the sheet.

Although 200 ppm of chloride is sufficient to induce SCC in a highly stressed metal, it is possible that chlorides were concentrated in the cracked region, perhaps by localized boiling or by concentration beneath deposits. It is worth noting in this regard that all of the cracking occurred at the hot end of the exchanger. This last observation may also reflect the temperature dependence of chloride SCC in austenitic stainless steels, which is not frequently reported at temperatures below 160°F (70°C).

Figure 9.8 Short, transverse fissure adjacent to the rolled end of the tube.

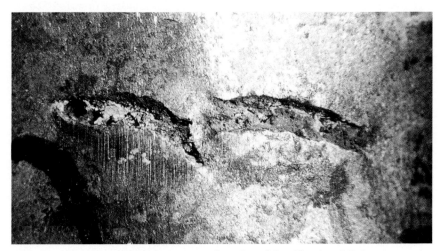

Figure 9.9 Fissures originating on external surface. (Magnification: 7.5×.)

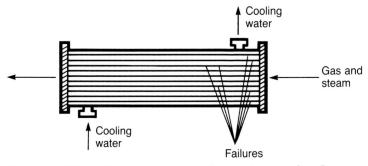

Figure 9.10 Schematic of heat exchanger showing location of cracks.

CASE HISTORY 9.2

Industry:	Chemical process
Specimen Location:	Surface condenser tube
Specimen Orientation:	Horizontal
Environment:	Internal: Treated cooling water, pH 7.0–7.5, chromate 10–13 ppm, chloride 0.2–0.4 ppm
	External: Steam and condensate at a partial vacuum of 4–10 in. (10–25 cm) of mercury
Time in Service:	17 years
Sample Specifications:	¾ in. (19 mm) outer diameter, admiralty brass

A routine inspection of the tube bundle during a plant outage revealed fine cracks of the type shown in Fig. 9.11. Scattered longitudinal cracks were observed along the lengths of most tubes. The external surface was covered with a thin film of black copper oxide and deposits. The bundle had been exposed to ammonia levels that produced ½ ppm of ammonia in the accumulated condensate.

Microstructural analysis revealed branched cracks running across the metal grains (transgranular) originating on the external surface.

The longitudinal orientation of these cracks reveals that hoop (circumferential) stresses caused by internal pressurization provided the necessary stresses. Ammonia was the specific corrodent involved.

Note that the ½ ppm of ammonia measured in the accumulated condensate may not accurately reflect the potentially higher level of ammonia dissolved in a thin film of moisture that could form on the metal wall in the region of initial condensation. This film of water could be saturated with ammonia, increasing susceptibility of the metal to SCC.

Figure 9.11 A network of longitudinal cracks along the external surface. (Magnification: 7.5×.)

CASE HISTORY 9.3

Industry:	Chemical process
Specimen Location:	Tube from a heat exchanger—instrument air cooler
Specimen Orientation:	Horizontal
Environment:	Internal: Air entering the exchanger at 115°F (45°C)
	External: Treated cooling water at 50°F (10°C)
Time in Service:	4 months
Sample Specifications:	⅝ in. (16 mm) outer diameter, admiralty brass

Cracks of the type illustrated in Fig. 9.12 occurred in many tubes of this new exchanger. All cracks occurred at the air-entry end of the cooler (Fig. 9.13) and had the brittle appearance typical of stress-corrosion cracks.

Close examinations of the internal surface under a low-power stereoscopic microscope revealed—in addition to the open cracks illustrated—numerous tight secondary cracks. Tube surfaces were smooth, uncorroded, and free of deposits.

Microstructural examinations revealed that the tube was annealed brass. All cracks originated on the internal surface at pit sites. Cracks were fine, branched, and ran through the metal grains (transgranular). Crack density over the specimens examined averaged 40 cracks per linear inch (16 cracks/cm) of surface.

The transverse orientation of the cracks, coupled with their predominance along one side, reveals that bending of the tube induced the responsible stresses. The rapidity of crack propagation and the high crack density indicate that the stress level was high.

The confinement of the cracks to a specific area of the cooler suggests that condensate from atmospheric moisture initially formed in this area and dissolved a corrodent from the atmosphere such as ammonia, sulfur dioxide, or oxides of nitrogen. Since the previous cooler had been in service for 20 years, it is conjectured that the rapid failure of this exchanger was due principally to very high bending stresses, which may have been induced during construction of the cooler.

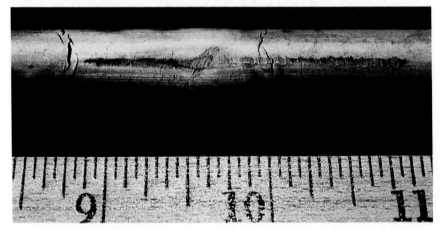

Figure 9.12   Discontinuous, transverse cracks; close examination of the photograph will disclose at least five tight, secondary cracks.

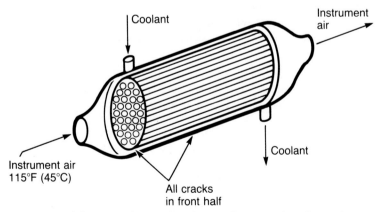

Figure 9.13   Schematic of air cooler showing location of cracks at the air-entry end.

CASE HISTORY 9.4

Industry:	Chemical process
Specimen Location:	Heat exchanger tube from a carbon dioxide compressor intercooler
Specimen Orientation:	Horizontal
Environment:	Internal: Carbon dioxide temperature 280–163°F (138–73°C), pressure 1735 psi (12 MPa)
	External: Treated cooling water, phosphate and zinc and a biodispersant, temperature 104°F (40°C), pressure 64 psi (441 kPa), pH 8–9, chloride 200 ppm
Time in Service:	2 years
Sample Specifications:	$^9/_{16}$ in. (1.4 cm) outside diameter, 304 stainless steel tube

A total of 13 tubes in the exchanger had suffered cracking of the type illustrated in Fig. 9.14. The cracks are predominantly longitudinally oriented.

Microstructural examinations revealed that the cracks originated on the external surface (Fig. 9.15). The cracks were highly branched and transgranular. The branched, transgranular character of these cracks is typical of stress-corrosion cracking of austenitic stainless steels. The thick-walled fracture faces are also typical of cracking by this mode.

The longitudinal orientation of the cracks reveals that the stresses responsible for the cracking were hoop stresses (circumferentially oriented) resulting from the internal pressure. The probable specific cracking agent is chlorides, which may have concentrated at possible boiling sites on the external surface. Note that internal and external operating conditions and environments, although physically separated by the tube wall, each contributed a necessary condition to induce stress-corrosion cracking.

Remedial measures include reduction or elimination of chlorides or replacement of the 304 stainless steel with a metal that is resistant to chloride stress-corrosion cracking.

Figure 9.14   Cracks on external surface.

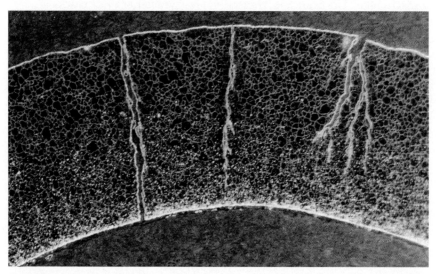

Figure 9.15   Cracks originating on the external surface. Note the branching. (Magnification: 33×.)

CASE HISTORY 9.5

Industry:	Chemical process—plastics
Specimen Location:	Coolant piping connections in extrusion barrel cooling jacket
Specimen Orientation:	Vertical
Environment:	Internal: Softened cooling water, temperature 250°F (120°C), pressure 25 psi (172 kPa), pH 7.8–8.5
	External: Ambient
Time in Service:	4 months
Sample Specifications:	½ inch (1.3 cm), outer diameter, 304 stainless steel tube

Figure 9.16 depicts the arrangement of connections in the extrusion barrel cooling jacket. Embrittlement and cracking of these stainless steel connections (Figs. 9.17 and 9.18) were occurring as frequently as every 4 to 6 weeks.

Close visual inspection of the blunt, fractured ends revealed small secondary cracks along the internal surface. Internal surfaces were covered with a layer of deposit that was highly alkaline (Fig. 9.18).

Microstructural examinations disclosed highly branched, predominantly transgranular cracks originating on the internal surface. Cracks of this form are typical of SCC in austenitic stainless steels.

The transverse (circumferential) crack path reveals that the stresses responsible for SCC were axially oriented; that is, the tube was pulled at its ends. Residual tube-forming stresses may also have contributed in this case. The specific cracking agent was caustic, which was apparently concentrated by evaporation when water flashed to steam in these locations.

Several approaches are available to eliminate this problem. If stresses cannot be sufficiently reduced, a metal possessing greater resistance to caustic SCC can be specified for replacement tubing. Alteration to the environment, such as eliminating alkalinity or increasing system pressure to prevent flashing (thereby minimizing the potential for concentration of caustic), would also prevent failure by this mechanism.

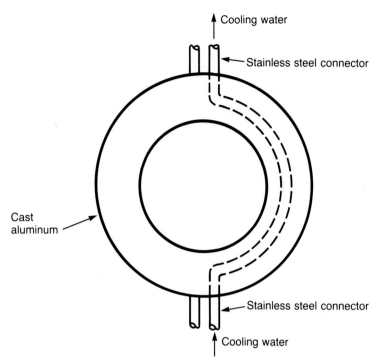

Figure 9.16   End view of plastics extrusion barrel cooling jacket.

Figure 9.17   Stainless steel connectors showing circumferential fractures.

Figure 9.18 Cracked connectors; note brittle, thick-walled character of the cracked faces and the accumulation of alkaline deposits on the internal surface of the connector on the left.

CASE HISTORY 9.6

Industry:	Chemical process
Specimen Location:	Orthodichlorobenzene (ODCB) column condenser tube
Specimen Orientation:	Vertical
Environment:	Internal: Process Stream, ODCB 99.4%, $COCl_2$ 0.5%, HCl 0.1%; entry temperature 220°F (105°C); exit temperature 175°F (80°C)
	External: Treated cooling water entry temperature 80°F (27°C), exit temperature 130–212°F (55–100°C)
Time in Service:	3 months
Sample Specifications:	1½ in. (3.8 cm) outside diameter, 304 stainless steel tube

Numerous tubes in this new heat exchanger had suffered circumferential cracks of the type illustrated in Fig. 9.19. All cracking had occurred in the top of this vertical condenser, within 1 in. (2.5 cm) of the rolled areas in the vapor space (Fig. 9.20).

Microstructural examinations revealed branched, transgranular cracks originating on the external surface (treated cooling water). Analysis of material covering the crack surfaces revealed the presence of chlorine.

The visual and microscopic appearance of the cracks, coupled with the presence of chlorine-containing corrosion products on the cracked surfaces, identifies this failure as chloride SCC. The circumferential orientation of the cracks and their location on just one side of the tube near its entrance into the tube sheet reveal that the stresses responsible for the cracking were bending stresses concentrated near the plane of constraint (tube sheet).

Several factors appeared to have contributed to the delivery of chloride to the stressed area. First, early in the short life of the exchanger the cooling water was contaminated with hydrochloric acid, resulting in a depression of pH to 4. Second, the valve regulating cooling water flow into the condenser was almost closed (open 20%), resulting in low flow of cooling water. Consequently, at times the cooling water would exit the condenser at the boiling point. The vent provided to discharge normal quantities of vapors from the vapor space was apparently incapable of accommodating the large quantities of vapors produced by the boiling. Water droplets in the vapor space dissolved hydrochloric acid vapors. Chloride was then delivered to the stressed tubes by water droplets, first impinging and then evaporating on the tube surfaces.

Figure 9.19   Circumferential cracks along one side of the heat exchanger tube. Note proximity of cracks to the plane of the tube sheet, as indicated by the tube sheet grooves (left). Expansion of the cracks may have occurred during removal of the tube from the exchanger.

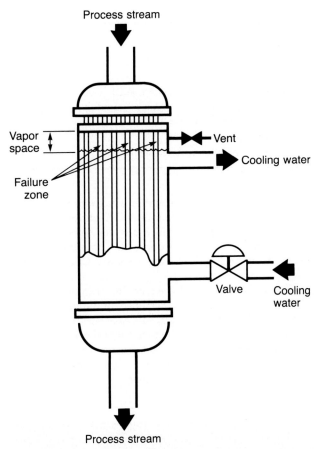

Figure 9.20   Schematic representation of vertical heat exchanger showing location of cracks in the vapor space.

CASE HISTORY 9.7

Industry:	Power generation
Specimen Location:	Heat exchanger tube from chiller
Specimen Orientation:	Horizontal
Environment:	Internal: Untreated water at 42°F (6°C)
	External: Lithium bromide solution at 40°F (4°C) and a partial pressure of 4 mm of mercury
Time in Service:	10 years
Sample Specifications:	¾ in. (19 mm) outer diameter, copper

As part of a protective maintenance activity, the chiller was examined by eddy current. An eddy current check revealed no failures. However, when the chiller was returned to service, leakage of chill water was detected. The unit was shut down, drained, and visually examined. Numerous branched cracks of the type shown in Fig. 9.21 were observed. Cracks were oriented longitudinally.

Microstructural examinations revealed that the cracking originated on the external surface, which was very smooth, uncorroded, and covered with a thin film of copper oxide. The SCC agent in this case was probably ammonia coupled with moisture and oxygen. The longitudinal orientation of the cracks reveals that the stresses responsible for the cracking were hoop (circumferential) stresses, apparently present during operation of the system. The unit was in service only during the summer.

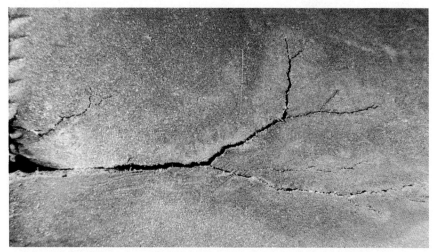

Figure 9.21   Branched, longitudinal cracks on the external surface. (Magnification: 7.5×.)

CASE HISTORY 9.8

Industry:	Steel
Specimen Location:	Condenser from turbo blower
Specimen Orientation:	Horizontal
Environment:	Internal: Lake water at 32–75°F (0–24°C), pH 8.2, chloride 0.3 ppm, 30 psi (207 kPa)
	External: Treated steam and condensate under a vacuum of 1–3 in. (25–76 mm) of mercury, condensate pH 8.6, temperature 78–102°F (26–39°C)
Time in Service:	27 years
Sample Specifications:	1 in. (25 mm) outer diameter, admiralty brass

Numerous failures of the type illustrated in Fig. 9.22 had occurred over a period of months. External surfaces are visually smooth, uncorroded, and covered with a thin layer of deposit. Internal surfaces are covered with a coating of copper carbonate.

Microstructural examinations revealed that branched cracks originated at shallow pit sites on the external surface. The pits, which may have formed during idle periods from differential oxygen concentration cells formed beneath deposits, acted as stress concentrators. The transverse (circumferential) crack orientation and the localization of cracks along just one side of the tube revealed that bending of the tube was responsible for the stresses involved.

Steam condensed in a thin film along the cold tube surface. This thin film of condensate then dissolved ammonia and oxygen present in the steam, which, in combination with stress, produced the observed cracking.

Figure 9.22   Transverse cracks originating on the external surface.

Introduction to Failure Modes Involving Mechanical Damage

Corrosion fatigue, cavitation, erosion, and erosion-corrosion differ from other forms of deterioration in the importance that mechanical factors play in the deterioration process. Deterioration by fatigue, erosion, and cavitation does not require a contribution from corrosion reactions. Rather, damage can be simply the consequence of purely mechanical interactions between the operating system and the metal. Repeated stressing of a metal above a threshold stress level can produce metal fatigue. Rapid, relative movement between a metal surface and a fluid can produce erosion. Rapid, cyclic, localized pressure changes within a fluid can produce short-lived vapor bubbles whose collapse (cavitation) can result in metal surface wastage (cavitation damage). Fatigue, erosion, and cavitation do not require a contribution by corrosion; but in typical cooling water systems, mechanical factors and corrosion reactions commonly combine synergistically to produce accelerated deterioration. Since the synergistic interactions between mechanical factors and corrosion is the rule rather than the exception in typical industrial environments, the discussion and case histories for Chap. 11, "Erosion-Corrosion," and Chap. 12, "Cavitation Damage," will reflect the importance of corrosion in deterioration involving mechanical phenomena.

The powerful influence of mechanical factors on these phenomena produces distinctive physical features on affected metal surfaces, as well as determining the locations where the damage occurs. Hence, metal wastage influenced by mechanical factors can be sensitive to geometric shapes and topographical features (surface contours). It is, in this sense, location specific.

Corrosion Fatigue

General Description

Corrosion fatigue, first mentioned in the literature in 1917, refers to the generation of cracks resulting from the combined effects of cyclic stresses and corrosion (Fig. 10.1). A near relative of fatigue, which results simply from cyclic stresses, corrosion fatigue differs from fatigue due to a synergism produced between the stresses and corrosion. Damage produced by the combined effects of the stresses and corrosion is significantly greater than the total damage produced by stresses and corrosion separately.

Several theories have been proposed to explain the corrosion-fatigue phenomena. One is that cyclic stressing causes repeated rupture of protective coatings. Corrosion-fatigue cracks propagate as the coating is successively reformed and ruptured along a plane.

No common industrial metal is immune to corrosion fatigue since some reduction of the metal's resistance to cyclic stressing is observed if the metal is corroded, even mildly, by the environment in which the stressing occurs. Corrosion fatigue produces fine-to-broad cracks with little or no branching. They are typically filled with dense corrosion product. The cracks may occur singly but commonly appear as families of parallel cracks (Fig. 10.2). They are frequently associated with pits, grooves, or other forms of stress concentrators. Like other forms of

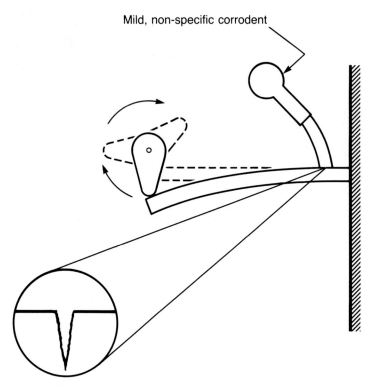

Figure 10.1   Schematic representation of a corrosion-fatigue crack propagating under the influence of a cyclic bending stress and exposure to a mild, nonspecific corrodent.

environmentally induced cracking, corrosion fatigue produces thick-walled fractures, even in ductile metals (Fig. 10.3).

Locations

Common locations of corrosion fatigue are planes of constraint, such as tube sheets, where cyclic stresses may be intensified. See Case History 10.1. Under conditions of flow-induced vibration in heat exchanger tubes, corrosion-fatigue cracking at midspan regions has been reported. Corrosion-fatigue crack initiation at pit sites is common and can be encountered where sufficient cyclic stresses are operating in pitted regions.

Critical Factors

Numerous factors can have a potentially significant effect on corrosion-fatigue cracking. Most of these relate to stress and the corrosiveness of

Figure 10.2 A family of short, transverse corrosion-fatigue cracks originating on the external surface.

the environment. The effect of these factors varies with the metals and environments involved.

Stress

The level of stress may be, and generally is, much less than the yield strength of the metal. However, in general, higher stresses increase crack growth rate and the number of cracks initiated.

Perhaps the most important stress factor affecting corrosion fatigue is the frequency of the cyclic stress. Since corrosion is an essential component of the failure mechanism and since corrosion processes typically require time for the interaction between the metal and its environment, the corrosion-fatigue life of a metal depends on the frequency of the cyclic stress. Relatively low-stress frequencies permit adequate time for corrosion to occur; high-stress frequencies may not allow sufficient time for the corrosion processes necessary for corrosion

Figure 10.3 Blunt fracture edges typical of corrosion-fatigue cracks.

fatigue. As stress frequency increases, the failure mode gradually shifts from corrosion fatigue to simple fatigue at very high frequencies. A fatigue limit, which is characteristic of simple fatigue, does not typically exist where corrosion fatigue is the failure mechanism. (Fatigue limit is the maximum stress below which a metal can endure an infinite number of stress cycles.)

Corrosion

In general, as the corrosiveness of the environment increases, the rate of crack growth also increases. Environmental factors constituting "corrosiveness" vary with the metal under consideration.

An important corrosion factor is the formation of crack initiation sites by localized corrosion, such as pitting. Such sites serve as stress concentrators. Stress concentrators locally elevate the stress level due to the geometry of the site (Fig. 10.4). It is common to find corrosion-fatigue cracks growing from such sites because locally elevated stresses are sufficiently high to rupture protective coatings. In uncorroded areas the stress level is lower and may be insufficient to cause coating rupture. It is also quite possible that the local environment at a pit site is more corrosive due to the concentration of aggressive substances there and, therefore, more prone to produce corrosion fatigue in the presence of cyclic stress.

Identification

Corrosion fatigue produces tightly closed, relatively straight, typically unbranched cracks that are frequently observed in parallel families. The tightness of the cracks may make them difficult to detect visually. They are commonly found at the bottom of localized corrosion sites such as pits, although they can form in the absence of pits. They may be oriented longitudinally or transversely or skewed, depending on the orientation of the maximum cyclic tensile stress. The cracks will run perpendicularly to this stress. The cracks have a thick-walled, brittle appearance, even in ductile metals. Corrosion products commonly cover the fracture faces but may not be present on the surface of the component.

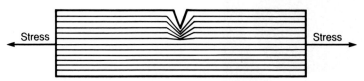

Figure 10.4 Schematic representation of localized increase in stress at a surface discontinuity.

Corrosion-fatigue cracks can be detected by nondestructive testing techniques such as magnetic particle inspection, radiography, ultrasonics, and dye penetrant. Corrosion-fatigue cracks may occur in numerous tubes simultaneously. Nondestructive testing of tubes at locations similar to those in which cracks are observed can be useful.

Elimination

Mitigation or elimination of corrosion-fatigue cracks involves gaining control of the critical factors that govern the mechanism.

Stress

Reduce or eliminate cyclic stresses. Reduction or elimination of cyclic stresses frequently requires equipment design modifications. Attention to external sources of cyclic stress, such as vibrations from machinery and water hammer, is also required. Internal sources of cyclic stress may include thermal expansion and contraction and flow-induced vibration. Cracking from flow-induced vibration is often restricted to peripheral rows of heat exchanger tubes. Reduction of flow-induced vibration requires correct spacing of tube supports and/or a change in the placement and design of baffles. A change of tube metallurgy without an appropriate alteration of tube supports and/or baffling may cause flow-induced vibration problems where none existed previously.

Corrosion

Techniques for controlling the corrosion factor can be categorized into three basic approaches:

1. *Alter the environment to render it less corrosive.* This approach may be as simple as maintaining clean metal surfaces. It is well known that the chemistry of the environment beneath deposits can become substantially different than that of the bulk environment. This difference can lead to localized, underdeposit corrosion (see Chap. 4, "Underdeposit Corrosion"). The pit sites produced may then induce corrosion fatigue when cyclic stresses are present. The specific steps taken to reduce corrosivity vary with the metal under consideration. In general, appropriate adjustments to pH and reduction or elimination of aggressive ions should be considered.

2. *Alter the metal to achieve greater corrosion resistance.* Since a metal's susceptibility to corrosion fatigue depends largely on its corrosion resistance in a particular environment, improving corrosion

resistance also improves resistance to corrosion fatigue. Proper selection of a metal may eliminate corrosion-fatigue problems altogether. However, care must be taken that the introduction of a new metal does not result in trading corrosion-fatigue cracking for some other form of environmentally sensitive cracking, such as stress-corrosion cracking.

3. *Separate the metal from the environment with a physical barrier.* Many corrosion inhibitors make use of this principal to protect metals. Proper use of an appropriate inhibitor may reduce or eliminate pitting. Pits are frequently initiation sites for corrosion-fatigue cracks. The effectiveness of inhibitors depends upon their application to clean metal surfaces. An example of this method is the use of zinc coatings on steel to stifle pit formation.

Cautions

Corrosion-fatigue cracks can be difficult to see because of their tightness. Nondestructive testing techniques should be used in suspect areas. Evidence of gross corrosion does not necessarily accompany corrosion-fatigue cracking. Deposits may cover crack sites, rendering them undetectable until the deposits are removed. Simple visual examinations of crack appearance may be insufficient to distinguish between the various modes of environmentally induced cracking such as corrosion fatigue, stress-corrosion cracking, and sometimes hydrogen embrittlement. In such cases a formal metallographic examination is required. Corrosion fatigue may be distinguished from simple fatigue by observing that fatigue cracks commonly occur singly and corrosion-fatigue cracks frequently occur in families.

Related Problems

See Chap. 9, "Stress-Corrosion Cracking."

CASE HISTORY 10.1

Industry:	Utility
Specimen Location:	Surface condenser tube, inlet end
Specimen Orientation:	Horizontal
Environment:	Internal: Circulating water pH 7.8–8.2
	External: Steam and condensate treated with an oxygen scavenger and ammonia, pH 7.8–9.2, temperature ambient to 150°F (66°C)
Time in Service:	25 years
Sample Specifications:	1 in. (25 mm) outer diameter, 30 ft (9 m) long, inhibited admiralty brass

Figure 10.5 shows the appearance of numerous failures that had occurred in a short time in the water box inlet end. Cracking of this type was a recurrent problem in this condenser. Approximately 9% of the tubes in the condenser had been plugged. The condenser was in cyclic service, although the failures had occurred while the boiler itself was out of service.

Note the proximity of the circumferential crack to the rolled section of tube where it had contacted the tube sheet. (Slight distortion of the tube occurred during its removal from the tube sheet.) Figure 10.6 shows the straight, unbranched character of the crack.

Microstructural examinations of tube wall cross sections taken from the cracked region revealed straight, unbranched cracks running directly across the metal grains (transgranular), typical of corrosion-fatigue cracks. The cracks originated on the external surface, although evidence of very shallow corrosion-fatigue cracks was also observed on the internal surface. Corrosion was superficial.

The cyclic stresses responsible for this failure were apparently bending stresses associated with cyclic thermal expansion and contraction.

Figure 10.5   Circumferential corrosion-fatigue crack adjacent to location of tube sheet.

Figure 10.6   Straight, unbranched crack on the surface. (Magnification: 33×.)

CASE HISTORY 10.2

Industry:	Utility
Specimen Location:	Turbine condenser tube
Specimen Orientation:	Horizontal
Environment:	Internal: Well water 95°F (35°C), pH 7.9, sulfate 900 ppm, chloride 330 ppm, molybdate water treatment
	External: Steam and condensate
Time in Service:	1 year, cyclic operation
Sample Specifications:	¾ in. (19 mm) outer diameter, admiralty brass

The short circumferential cracks apparent in Fig. 10.7 are precursors of the complete, brittle fractures shown in Fig. 10.3. Numerous fractures had occurred adjacent to the tube sheet over the past year.

Microstructural examinations revealed that the primary cracks originated on the internal surface at the base of shallow pit sites. Fewer cracks originated on the external surface, also at the base of shallow pit sites. The cracks are straight, transgranular and filled with dense corrosion products, typical of corrosion-fatigue cracks. The cyclic stresses responsible for this cracking were apparently bending stresses produced by thermal expansion and contraction.

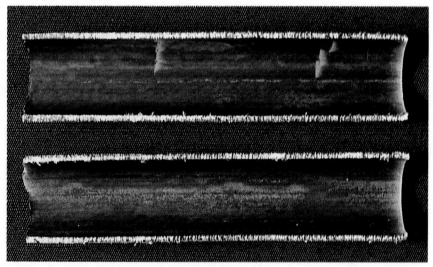

Figure 10.7   Short, tight, circumferential corrosion-fatigue cracks on the internal surface made apparent by discoloration from "weeping."

CASE HISTORY 10.3

Industry:	Refinery
Specimen Location:	Surface condenser: bed air blower cat cracker regeneration system
Specimen Orientation:	Horizontal
Environment:	Internal: Cooling water 90–120°F (32–49°C), pH 6.6–6.8 (pH 4 occasionally), 30–60 psi (0.2–0.4 MPa)
	External: 650 psi (4.5 MPa), steam condensate pH 8–9
Time in Service:	10 years
Sample Specifications:	¾ in. (19 mm) outer diameter, admiralty brass

Transverse cracks shown in Fig. 10.8 were found in a total of 15 tubes from random areas of the condenser. The equipment had been in continuous service for 2 years. Cracks had not been observed previously.

Visual examinations disclosed that the cracks originated on both the internal and external surfaces, although there were more on the internal surface. Thin coatings of deposits partially covered both the internal and external surfaces.

Microscopic examinations revealed that the cracks were unbranched and transgranular, typical of corrosion-fatigue cracks. These examinations also revealed that the cracks tended to initiate at sites of intergranular corrosion or at pit sites, both of which are stress concentrators. Intergranular corrosion and pitting are frequently associated with deposits covering the metal surface.

The orientation of the cracks reveals that cyclic bending stresses or cyclic axial stresses were active. The intensification of these stresses at pits and intergranular corrosion sites produced the cracks observed.

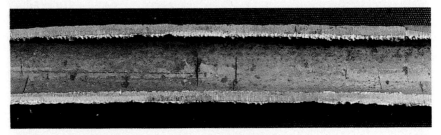

Figure 10.8   Transverse cracks on the internal surface.

CASE HISTORY 10.4

Industry:	Refinery
Specimen Location:	Surface condenser for air compressor
Specimen Orientation:	Horizontal
Environment:	Internal: Cooling water: phosphonate and zinc water treatment
	External: Steam and condensate with ammonia and hydrogen sulfide
Time in Service:	8 years
Sample Specifications:	¾ in. (1.9 cm) outer diameter, admiralty brass

Numerous transverse cracks of the type illustrated in Figs. 10.2 and 10.9 were discovered upon inspection. Cracks had not been observed previously.

Some cracks were clustered into small areas along one side of the tube, but close examination revealed numerous cracks scattered over the tube surface. Chemical spot tests revealed the presence of sulfur-containing compounds on the external surface.

Microscopic examinations revealed tight, unbranched, transgranular cracks originating on the external surface. Many of these cracks originated at shallow pockets of corrosion.

The orientation of the cracks indicates that cyclic bending stresses or cyclic axial stresses generated by thermal expansion and contraction provided the responsible stresses. The large number of crack initiation sites and the tightness of the cracks indicate high-level stresses.

Although reducing the corrosivity of the external environment would reduce cracking, in this case it is judged that cyclic, operational stresses are predominant. Until the source of the stresses is determined and reduced or eliminated, it is doubtful that a permanent solution can be achieved.

Figure 10.9   Close-up of cracks. (Magnification: 7.5×.)

CASE HISTORY 10.5

Industry:	Primary metals
Specimen Location:	Heat exchanger: aftercooler of an oxygen plant
Specimen Orientation:	Horizontal
Environment:	Internal: Oxygen
	External: Cooling water at 95°F (35°C), phosphate and zinc water treatment
Time in Service:	5 years
Sample Specifications:	⅝ in. (16 mm) outer diameter, copper

Longitudinal cracks of the type illustrated in Fig. 10.10 affected numerous tubes in the exchanger. Cracking was a recurrent problem.

Visual examinations disclosed several cracks, each very tight and difficult to see. Examinations under a low-power stereoscopic microscope revealed many short cracks running parallel to the primary crack.

Microstructural examinations revealed intergranular, sparsely branched cracks originating on the external surface. Some cracks initiated as transgranular fissures.

Intergranular corrosion-fatigue cracks in copper may be difficult to differentiate from stress-corrosion cracking. The longitudinal orientation of the cracks revealed that the cyclic stresses were induced by fluctuations in internal pressure.

Figure 10.10   Longitudinal corrosion-fatigue crack originating on the external surface.

Erosion-Corrosion

General Description

Erosion-corrosion can be defined as the accelerated degradation of a material resulting from the joint action of erosion and corrosion when the material is exposed to a rapidly moving fluid. Metal can be removed as solid particles of corrosion product or, in the case of severe erosion-corrosion, as dissolved ions.

Metal surfaces in a well-designed, well-operated cooling water system will establish an equilibrium with the environment by forming a coating of protective corrosion product. This covering effectively isolates the metal from the environment, thereby stifling additional corrosion. Any mechanical, chemical, or chemical and mechanical condition that affects the ability of the metal to form and maintain this protective coating can lead to metal deterioration. Erosion-corrosion is a classic example of a chemical and mechanical condition of this type. A typical sequence of events is:

1. A protective coating forms over the metal surface.
2. The coating is mechanically removed by the abrasive effects of a high-velocity fluid.
3. The protective coating reforms on the metal surface via a corrosion process.

Steps 2 and 3 may occur almost simultaneously.

Metals that depend on a relatively thick protective coating of corrosion product for corrosion resistance are frequently subject to erosion-corrosion. This is due to the poor adherence of these coatings relative to the thin films formed by the classical passive metals, such as stainless steel and titanium. Both stainless steel and titanium are relatively immune to erosion-corrosion in most cooling water environments.

Most metals are subject to erosion-corrosion in some specific environment. Soft metals, such as copper and some copper-base alloys, are especially susceptible. Erosion-corrosion is accelerated by, and frequently involves, a dilute dispersion of hard particles or gas bubbles entrained in the fluid.

When very high velocities are encountered, metal loss from erosion-corrosion can be general. Typically, however, erosion-corrosion produces localized metal loss in immediate proximity to the disrupted flow. Smooth, rolling, wavelike surface contours are often produced, or distinct, horseshoe-shaped depressions (Fig. 11.1) or comet tails

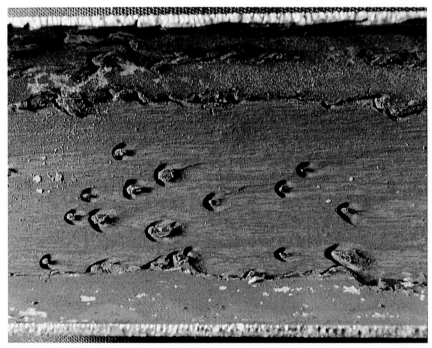

Figure 11.1 Horseshoe-shaped depressions on the internal surface of a brass heat exchanger tube.

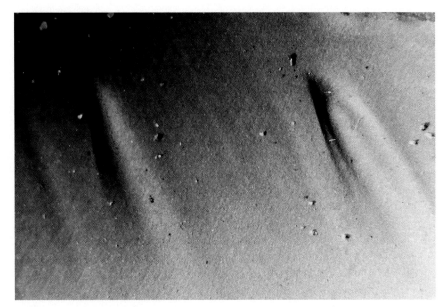

Figure 11.2 Comet-tail erosion patterns on copper. (Magnification: 15×.)

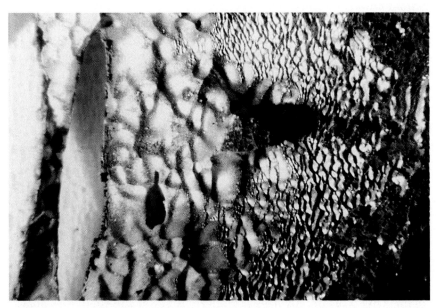

Figure 11.3 Sand dunelike erosion-corrosion patterns on the inlet end of a steel heat exchanger tube. (Magnification: 7×.)

Figure 11.4 Deposits covering intact metal surfaces at sites of horseshoe-shaped erosion-corrosion depressions. (Magnification: 15×.)

aligned in the flow direction are observed (Fig. 11.2). Sand dunelike contours are sometimes observed (Fig. 11.3).

Affected areas are essentially free of deposits and corrosion products, although these may be found nearby (Fig. 11.4). Affected areas may be covered with deposits and corrosion products if erosion-corrosion occurs intermittently, and the component is removed following a period in which erosion-corrosion was inactive.

Locations

Favored locations for erosion-corrosion are areas exposed to high-flow velocities or turbulence. Tees, bends, elbows (Fig. 11.5), pumps, valves (Fig. 11.6), and inlet and outlet tube ends of heat exchangers (Fig. 11.7) can be affected. Turbulence may be created downstream of crevices, ledges (Fig. 11.8), abrupt cross-section changes, deposits, corrosion products, and other obstructions that change laminar flow to turbulent flow.

Erosion-corrosion problems on the outside of tubes are frequently associated with impingement of wet, high-velocity gases such as steam. This typically involves peripheral tubes at the shell inlet nozzle (Fig. 11.9). Baffle and tube interfaces may also be affected.

Figure 11.5   Erosion-corrosion at elbow of a brass tube. Note also the horseshoe-shaped depressions and comet tails aligned with flow direction in the straight section.

Critical Factors

Erosion-corrosion is a fairly complex failure mode influenced by both environmental factors and metal characteristics. Perhaps the most important environmental factor is velocity. A threshold velocity is often observed below which metal loss is negligible and above which metal loss increases as velocity increases. The threshold velocity varies with metal and environment combinations and other factors.

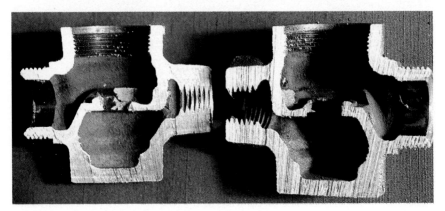

Figure 11.6   Split valve body showing metal loss at the valve seat.

Figure 11.7   Metal loss on the internal surface at an inlet end (equally spaced circumferential gouges near the tube end were created during tube removal).

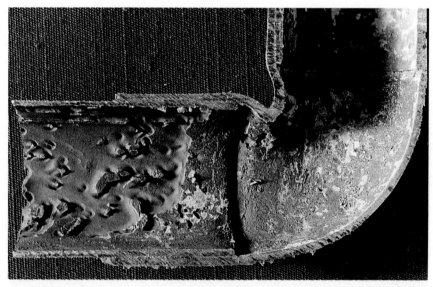

Figure 11.8   Turbulence created by the ledge at the tube end caused erosion-corrosion
immediately downstream.

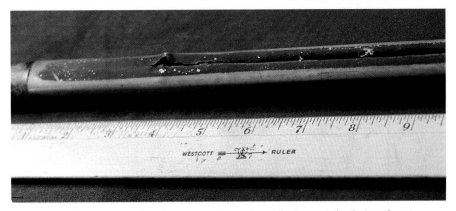

Figure 11.9 Erosion-corrosion damage on the surface of a brass tube facing the steam inlet nozzle.

Although it is entirely possible for erosion-corrosion to occur in the absence of entrained particulate, it is common to find erosion-corrosion accelerated by a dilute dispersion of fine particulate matter (sand, silt, gas bubbles) entrained in the fluid. The character of the particulate, and even the fluid itself, substantially influences the effect. Eight major characteristics are influential: particle shape, particle size, particle density, particle hardness, particle size distribution, angle of impact, impact velocity, and fluid viscosity.

In addition to fluid velocity, other characteristics of the eroding fluid can exert a marked influence on the erosion-corrosion process. Among the important factors are the following:

1. *Corrosivity.*   In general, the more corrosive the fluid, the greater the rate of metal loss. It is interesting to note that the fluid velocity can affect the apparent corrosiveness of a fluid. If aggressive substances, such as oxygen, carbon dioxide, and hydrogen sulfide in the case of steel, are present, higher velocities increase the supply rate of the substance to the metal surface, thereby increasing the corrosion rate. On the other hand, if corrosion inhibitors, such as passivating substances in the case of stainless steel, titanium, or aluminum, are present, higher velocities increase the supply rate of these inhibitors to the metal surface, resulting in a decrease in the corrosion rate.

2. *Temperature.*   In general, the higher the temperature, the higher the corrosion rate.

3. *pH.*   The effect here varies with the metal and environment combination. All other factors being equal, erosion-corrosion of mild steel

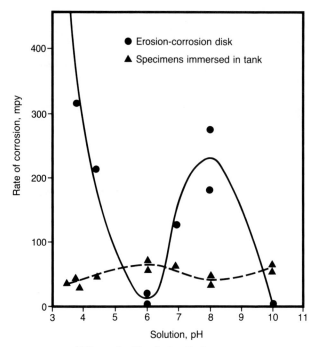

Figure 11.10 Effect of pH of distilled water on erosion-corrosion of carbon steel at 122°F (50°C) (velocity, 39 ft/s, 12 m/s). (SOURCE: *M. G. Fontana and N. D. Greene,* Corrosion Engineering, *2d ed., 1978, p. 75. Reprinted with permission from McGraw-Hill, Inc.*)

is minimized at a pH of 9.6 or higher. Note Fig. 11.10. Copper and copper-base alloys are adversely affected by high pH only if ammonia is present.

Impediments to water flow resulting from inadequate equipment design or lodgement of foreign objects in the tubes can exercise a dramatic effect on the erosion-corrosion process. Much of this influence is linked to the creation of turbulence and the simple increase in fluid velocity past obstructions. The importance of these factors is quickly recognized when the phenomenon of threshold velocity is considered.

For example, unsatisfactory equipment design has caused inlet-end erosion due to improper distribution of water in the water box. Poor design and construction practices have caused erosion-corrosion of the tube sheet due to unsatisfactory water distribution or leakage past pass partitions. Deep channels (wire drawing) or "worm holes" have been observed in tube holes due to leaking tube-to-tube sheet joints.

Figure 11.11 Erosion site resulting from lodgement of a foreign body inside a tube. (Magnification: 8×.)

A common cause of erosion is partial obstruction of tubes by foreign bodies. At the inlet end, for example, debris such as sticks, glass fragments, and wood chips may lodge in tube ends or be held against the tubes by water flow. The nominal velocity of the water past the obstruction increases according to the degree of obstruction. It can be shown

Figure 11.12 Erosion site resulting from lodgement of a foreign body inside a tube. (Magnification: 8×.)

Figure 11.13   Erosion site resulting from lodgement of a foreign body inside a tube. (Magnification: 8×.)

that if two-thirds of the tube opening is blocked, the water velocity past the obstacle can increase by 2½ times.

Some debris may enter the tube and then become lodged. Turbulence created at the site of the debris often produces crescent or irregularly shaped erosion-corrosion patterns (Figs. 11.11 through 11.13).

Identification

The classic signature of erosion-corrosion is the formation of horseshoe-shaped depressions, comet tails, grooves, or sand dunelike surface contours oriented along the direction of fluid flow (Figs. 11.1, 11.2, 11.3, 11.5, and 11.8). Occasionally, erosion-corrosion will produce smooth, almost featureless, surface contours (Fig. 11.15), although even in this case oriented metal loss often exists around the perimeter of the affected region. If erosion-corrosion has been recently active, affected surfaces will be free of accumulated deposits and corrosion products.

The early stages of impingement attack on peripheral tubes near the condenser inlet nozzle will give the external tube surfaces a polished appearance. As attack progresses, the tube surface becomes rough and jagged to the touch (Fig. 11.9).

NOTE: Nonferromagnetic (metals that cannot be magnetized, such as copper and brass) tubes that are not visually accessible can be examined nondestructively using eddy current testing.

Elimination

Due to the numerous factors that can contribute to an erosion-corrosion process, numerous approaches can mitigate or eliminate the problem.

1. Because erosion-corrosion is directly linked to velocity and turbulence, reducing velocity and turbulence are important first steps in elimination. Reducing velocity and turbulence often requires design changes such as the following:

- Increase tube or nozzle diameters.

- Increase bend radii.

- Eliminate or diminish flow disrupters.

- Alter the distribution pattern of incoming water by installing diffusers in the inlet region.

- For inlet or outlet end erosion-corrosion, either extend tube ends 3 or 4 inches into the water box or install sleeves, inserts, or ferrules into the tube ends. These should be a minimum of 5 inches long. The ferrules may be nonmetallic or erosion-resistant metals, such as stainless steel, if galvanically compatible. The end of the ferrule should be feathered to prevent turbulence.

Deterioration of tubes near the inlet nozzle of condenser shells due to impingement of water and steam mixtures can be alleviated through the use of appropriately placed and sized baffles or impact plates or by applying clip-on impingement shields to the tubes (see Case History 11.3).

2. Because alterations to equipment design can be cumbersome and expensive, a more economical approach may be to change the metallurgy of affected components. Metals used in typical cooling water environments vary in their resistance to erosion-corrosion. Listed in approximate order of increasing resistance to erosion-corrosion, these are copper, brass, aluminum brass, cupronickel, steel, low-chromium steel, stainless steel, and titanium.

The resistance of a metal to erosion-corrosion is based principally on the tenacity of the coating of corrosion products it forms in the environment to which it is exposed. Zinc (brasses), aluminum (aluminum brass), and nickel (cupronickel) alloyed with copper increase the coating's tenacity. An addition of ½ to 1¼% iron to cupronickel can greatly increase its erosion-corrosion resistance for the same reason. Similarly, chromium added to iron-base alloys and molybdenum added to austenitic stainless steels will increase resistance to erosion-corrosion.

Occasionally, tubes affected by erosion-corrosion or erosion processes may be confined to specific regions of the equipment. In this case, only the metallurgy of affected tubes needs to be altered. For example, impingement attack of peripheral tubes at the inlet nozzle of a condenser can be alleviated simply by specifying more erosion-resistant metals, such as 70:30 cupronickel, stainless steel, or titanium, for the affected tubes. (In this regard, see the note under "Cautions" relative to tube vibration.)

3. Entrainment of fine particulate matter such as sand and silt in cooling water can contribute significantly to erosion-corrosion. In these cases it is important to eliminate or reduce the amount of particulate by settling or filtration. It may also be necessary to reduce or eliminate entrained gas bubbles.

4. Significant erosion-corrosion is also associated with lodgement of large foreign objects (sticks, stones, shell fragments) on the face of the tube sheet or within tubes. As a source of erosion-corrosion, this condition has probably been underemphasized. Reduction or elimination requires frequent cleaning of screens and filters and periodic cleaning of the exchanger.

5. Corrosivity of the water has a direct influence on erosion-corrosion susceptibility. Opportunities for reduction of corrosivity are often limited, but the following are offered as guidelines:

- Eliminate or reduce the amount of corrosive ions.
- Decrease water temperature.
- Appropriately adjust pH.

6. Use of inhibitors. Because corrosion is such a vital aspect of the erosion-corrosion process, inhibitors that will reduce corrosion under conditions of high fluid velocity have been a cost-effective method of dealing with erosion-corrosion. For example, injection of ferrous sulfate either intermittently or continuously has been successful in inhibiting erosion-corrosion, especially with copper-base alloys.

7. Cathodic protection can be useful, although its ability to protect tube interiors is generally limited to the first 4 to 6 in. of tube length. Such systems, however, must be properly designed and maintained to be effective. Corrosion can be intensified if the polarity of the cathodic protection system is inadvertently reversed.

Cautions

A classic feature of erosion-corrosion is the directional character of the metal loss. The metal loss will be oriented along the direction of fluid flow or according to turbulence patterns. However, other corrosion modes may produce directionality in metal loss and could be confused with erosion-corrosion. For example, attack of steel by an acid may produce directional metal loss along mandrel marks and along machined surfaces. A corrosive fluid dripping or running down in response to gravity may also yield oriented metal loss.

When combating erosion-corrosion by changing tube metallurgy, caution must be exercised to ensure appropriate tube and baffle spacing so that vibration-associated cracking problems are not introduced.

Corrosion tests of metals under static conditions reveal nothing relating to erosion-corrosion susceptibilities. It is entirely possible that a metal tested under static conditions will fail in service when sufficient fluid velocity produces erosion-corrosion. Similarly, it has been observed that galvanic corrosion between coupled, dissimilar metals may be accelerated or even initiated under flow conditions when little or no galvanic corrosion is observed under static conditions (see Chap. 16, "Galvanic Corrosion").

CASE HISTORY 11.1

Industry:	Utility
Specimen Location:	Surface condenser tube, main condenser bundle
Specimen Orientation:	Horizontal
Environment:	Internal: Untreated river water, pH 8.0–8.2, total dissolved solids 300 ppm, M alkalinity 100 ppm
	External: Steam and condensate, pH 9.0, ammonia 300 ppb, oxygen 20 ppb, conductivity 3.0 μmhos/cm
Time in Service:	25 years
Sample Specifications:	¾ in. (19 mm) outside diameter, brass

Metal loss of the type illustrated in Fig. 11.14 occurred on the internal surface at the midsection of the tube. Note the erosion grooves oriented in the direction of flow. Metal loss at the inlet end was much more severe and had produced a smooth, relatively featureless contour (Fig. 11.15). Eroded areas were free of corrosion products and deposits.

Generally, maximum flow velocities for brass are between 3 and 5 ft/s (0.9 and 1.5 m/s) depending on environmental conditions. In this case, sand and silt entrained in the river water contributed to the erosiveness of the fluid.

Figure 11.14 Contour of the internal surface at midspan.

Figure 11.15 Relatively smooth, lightly rolling contour of the internal surface at the inlet end.

CASE HISTORY 11.2

Industry:	Utility
Specimen Location:	Condenser tube, inlet end
Specimen Orientation:	Horizontal
Environment:	Internal: Cooling water, 60–100°F (16–38°C), pH 7.8–8.4, conductivity 800–3000 μmhos/cm
	External: Steam and condensate, pH 8.2–9.2
Time in Service:	6 years
Sample Specifications:	1 in. (25 mm) outside diameter, 90:10 cupronickel

Figures 11.11 through 11.13 illustrate the type of metal loss responsible for a dozen tube leaks over a 3-month period. Note the highly localized character of the metal loss, as well as the unusual shapes and random orientations of the eroded sites. The sites were located at the inlet end and resulted from the lodgment of debris at the mouth of the tube. Highly localized turbulence created by this debris caused the erosion.

Visually, the sites resemble mechanically induced gouges or indentions in the tube wall. However, examinations of the microstructure at these sites revealed no distortion of the metal, which would certainly occur had the indentions been mechanically induced. The erosive character of the highly localized turbulent flow was the predominant aspect responsible for the metal loss, there being little or perhaps no contribution from corrosion of the metal.

CASE HISTORY 11.3

Industry:	Utility
Specimen Location:	Condenser tube, top row immediately adjacent to steam inlet
Specimen Orientation:	Horizontal
Environment:	Internal: Untreated river water
	External: Wet steam, pH 9.0–9.6
Time in Service:	17 years
Sample Specifications:	½ in. (13 mm) outside diameter, admiralty brass

The split apparent in Fig. 11.9 was located along the top of the tube facing the steam inlet nozzle. This is one of several tubes in this area having similar longitudinal splits. Leakage of river water from these tubes resulted in feedwater contamination, which turned out to be a major factor in tube failures in the boiler.

The split was centered in a distinct zone of severe wall thinning from the external surface. Close examinations of this surface under a low-power microscope revealed a forest of small, well-formed brass cones pointing straight up from the surface (Fig. 11.16). The external surface on the opposite side retained its smooth, original surface.

The metal loss and resulting split were caused by erosion from impingement of high-velocity steam in which droplets of water were entrained. The conical pinnacles in the affected region are typical of deterioration by this mechanism.

Although the coolant (river water) was at relatively low pressure, measurements revealed a residual hoop stress in the tube of approximately 9000 psi (62 MPa). The longitudinal rupture occurred as a result of these stresses after erosion had sufficiently reduced wall thickness.

Elimination of this problem can be effected by one or more of several methods:

1. Separate moisture from the steam before it enters the condenser.

2. Specify metals that are more resistant to erosion, such as cupronickel, monel, and stainless steels. Only affected tubes need to be replaced by these metals.

3. Install a deflector shield to protect the area experiencing erosion. Note that shields could be clipped to the tubes themselves.

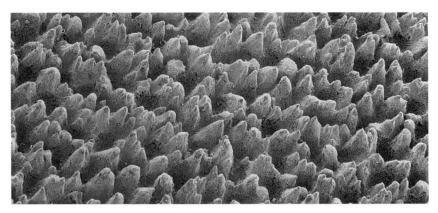

Figure 11.16 Scanning electron micrograph showing cones of brass created by impingement of high-velocity wet steam. (Magnification: 50×.)

CASE HISTORY 11.4

Industry:	Utility
Specimen Location:	Steam surface condenser
Specimen Orientation:	Horizontal
Environment:	Internal: River water treated with microbicides, 35–115°F (2–46°C), pH 8.2
	External: Steam and condensate, pH 8.5–9.0
Time in Service:	25 years
Sample Specifications:	⁷⁄₈ in. (22 mm) outside diameter, admiralty brass

Metal loss from the internal surface of the types illustrated in Figs. 11.17 and 11.18 had affected approximately 45 tubes over the previous 4 months. It was noted that metal loss along the bottom half of the tubes was more severe. Significant numbers of failures of this type had not been experienced before, although it was known that entrainment of silt in the cooling water occurred seasonally.

In this case, reduced river water levels resulting from a drought may have caused an increase in the amount of sand and silt entrained in the water, thereby increasing its erosive character. This may also account for the greater metal loss along the bottom half of the tubes, where heavy solids would tend to accumulate.

Note the incomplete attack of the surface in Fig. 11.17. This illustrates erosion-corrosion in its early stages. Compare this to the general metal loss illustrated in Fig. 11.18. Note the horseshoe-shaped depressions aligned with water flow. This tube illustrates erosion-corrosion in its mature form.

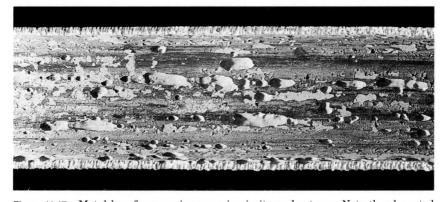

Figure 11.17   Metal loss from erosion-corrosion in its early stages. Note the elongated depressions aligned with water flow.

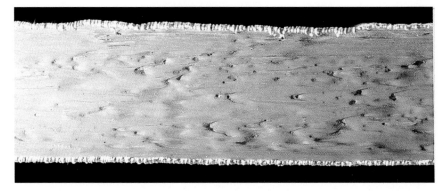

Figure 11.18   Erosion-corrosion in its mature form. Note the horseshoe-shaped depressions aligned with flow direction.

CASE HISTORY 11.5

Industry:	Utility
Specimen Location:	Surface condenser
Specimen Orientation:	Horizontal
Environment:	Internal: Brackish water treated with tolyltriazole and a chlorine biocide
	External: Steam and condensate
Time in Service:	2 years
Sample Specifications:	1 in. (2.5 cm) outside diameter, 90:10 cupronickel tubes

The first set of tubes received, which had been cleaned before submission for examination, contained randomly distributed, pinhole perforations at the bottom of large pit sites (Fig. 11.19). Associated with these sites was an irregular, arrowhead-shaped region of shallow corrosion. This region was covered with a corrosion product consisting of sparkling cuprous oxide crystals. These regions were immediately upstream of the pit site. Eddy current tests performed earlier in the year indicated that pitting had not occurred at that time.

At this point in the investigation, the relationship between the pits and the arrowhead-shaped regions of corrosion was uncertain. Several possible causes for the pitting were considered, such as siphonic gas exsolution, biological and/or microbiological activity, and debris (concrete chips, etc.) lodged in the tubes, but each was tentatively dismissed as improbable since none of the proposed mechanisms adequately accounted for all observations.

A few months later, a second set of tubes was submitted for examination. These tubes had not been cleaned. Close examination revealed arrowhead-shaped mounds of fibrous debris lodged on the tube wall (Fig. 11.20). Dislodgement of these mounds revealed an arrowhead-shaped region of shallow corrosion containing sparkling crystals of cuprous oxide, essentially identical to that described above.

The fibrous strands that composed the mounds adhering to the tube wall were examined and identified as seed hairs from grass. Subsequently, fibrous material was collected from the cooling tower basin and seed pods from grass growing in a bog located adjacent to and upwind of the cooling towers. This material was also examined and found to be identical to the seed hairs that composed the mounds adhering to the tube wall.

At this point, the sequence of events eventually leading to the tube wall perforations could be established:

1. Seed hairs from the grass formed, matured, and were carried by the wind into the basin of the cooling tower.
2. They collected in the cooling tower basin and were eventually carried into the cooling water system where their fibrous character permitted them to adhere to rough spots on the tube walls.
3. These spots then became localized collection sites for seed hairs, silt, and corrosion products, forming a mound that was shaped into an arrowhead by water flow.
4. The mound shielded the tube wall, causing underdeposit corrosion (see Chap. 4, "Underdeposit Corrosion").
5. When a critical mound size was attained, turbulence was created immediately downstream of the mound, resulting in erosion and eventual perforation of the tube wall.

The eddy current tests performed earlier detected no pits because pit formation depended on the contamination of the system by seed hairs that occurred subsequently to the eddy current testing.

This analysis underscores the importance of examining failed components before they are cleaned or in any way altered. It also demonstrates the potential complexity of failure analysis and the need that exists to discard explanations that do not adequately account for all relevant observations. Important also to note is the potential connectedness of environmental factors, such that the seasonal development of seed hairs in a field of grass near a cooling tower would eventually contribute to perforations of tubes in a condenser.

Figure 11.19 Perforated pit downstream of an arrowhead-shaped area of corrosion. (Magnification: 7.5×.)

Figure 11.20 Elongated mounds of fibrous debris attached to the internal surface. (Magnification: 7.5×.)

CASE HISTORY 11.6

Industry:	Utility
Specimen Location:	Condenser tube
Specimen Orientation:	Horizontal
Environment:	Internal: Treated cooling water adjusted with sulfuric acid for pH control and sodium hypochlorite added as a biocide; pressure 50 psi (345 kPa), temperature 100–120°F (38–49°C), water velocity 7 ft/s (2.1 m/s), pH 8.0–8.4, sulfate 500–1000 ppm, chloride 100–450 ppm, total hardness 500 ppm
	External: Steam and condensate
Time in Service:	4 years
Sample Specifications:	1 in. (2.5 cm) outside diameter, 304 stainless steel tubes

Severe, highly localized gouging of the internal surface of the type shown in Fig. 11.21 perforated a total of 100 tubes over a 4-year period. Visual inspections disclosed numerous additional gouged but, as yet, unperforated tubes. The gouging was confined to the first 8 in. (20 cm) of the inlet end of the tubes.

The gouge sites had a bright metallic luster and various shapes (Figs. 11.22 and 11.23). Microstructural examinations of the gouged regions revealed that plastic deformation of the metal had not occurred.

The rust-colored concrete chips shown in Fig. 11.24 were removed from the tube ends. Inspection of the water box revealed large quantities of debris adhering to the tube sheet. The gouging was caused by the lodgement of this hard debris at the inlet end of the tubes. Intense turbulence by the lodged debris was sufficient to cause highly localized erosion.

Although not illustrated in Figs. 11.21 through 11.23, each erosion site was composed of two or more gouged areas. This is characteristic of erosion due to lodged debris since at least two points of contact with the tube wall are required for a particle to remain in place.

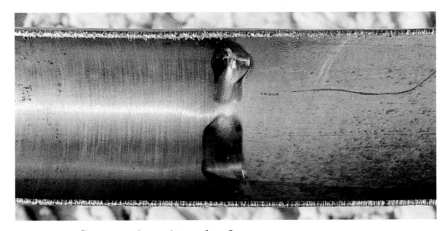

Figure 11.21   Severe gouging on internal surface.

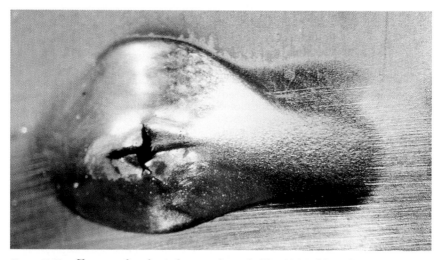

Figure 11.22   Close-up of perforated gouge shown in Fig. 11.21. (Magnification: 7.5×.)

Figure 11.23   Elliptical gouge on internal surface ½ in. from inlet end. The long axis parallels the longitudinal tube axis. (Magnification: 7.5×.)

Figure 11.24   Concrete chips covered with orange iron oxide that were removed from the tube ends.

CASE HISTORY 11.7

Industry:	Pulp and paper
Specimen Location:	Lubricating oil cooler
Specimen Orientation:	Horizontal
Environment:	Internal: Mill water, chlorinated (0.2 ppm residual), 120–160°F (49–71°C), pH 7.1–7.3, conductivity 100 μmhos/cm, Ryznar Stability Index 9.5–10.75
	External: Synthetic lubricating oil, 175°F (79°C)
Time in Service:	3 months to 3 years
Sample Specifications:	⅜ in. (10 mm) outside diameter, brass

The mill had experienced 30–40 heat exchanger failures over the previous 9 months. Service in these exchangers varied from 3 months to 3 years.

Close examinations revealed that metal loss was confined to ½ in. (13 mm) of the internal surface at the inlet end (Fig. 11.25). Note the two perforations near the bottom of the photograph. Generally, the metal loss was confined to the region within the tube sheet, but occasionally perforations occurred just beyond the tube sheet, resulting in leakage of cooling water.

Turbulence created at the inlet end resulted in erosion-corrosion. Problems of this type can be remedied by using appropriate inserts or ferrules.

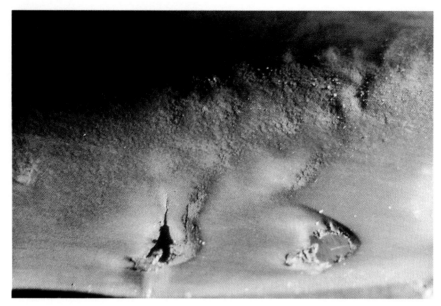

Figure 11.25   Erosion depressions and perforations at inlet ends of a brass tube.

CASE HISTORY 11.8

Industry:	Pulp and paper
Specimen Location:	Steam turbine surface condenser
Specimen Orientation:	Horizontal
Environment:	Internal: Brackish river water, pH 7.1
	External: Steam and condensate
Time in Service:	Unknown
Sample Specifications:	⅞ in. (22 mm) outside diameter, brass

Figure 11.7 illustrates the internal surface at the inlet end of the condenser. Approximately 2 in. (5 cm) of the surface is marked by mutually intersecting depressions and grooves. Areas of the internal surface downstream of this zone are smooth and covered with a thin layer of deposits. This typical case of inlet-end erosion can be eliminated by the techniques discussed earlier in this chapter under "Elimination."

CASE HISTORY 11.9

Industry:	Office building
Specimen Location:	Condenser, air-conditioning system
Environment:	Internal: Treated cooling water
	External: Freon refrigerant*
Time in Service:	13 years
Sample Specifications:	¾ in. (19 mm) outside diameter, copper tube; 2½ in. (6.3 cm) outside diameter, copper canister

Figure 11.26 shows a component removed from an air-conditioning compressor. Pinhole perforations in this component had allowed cooling water to leak into the freon. Many failures of this type had occurred previously. Examinations of the internal surfaces of both the canister and the tubes entering it revealed evidence of metal loss. Tiny perforations at the bases of deep grooves were noted in the tubes. Deep, general, smooth metal loss surrounded irregular islands of intact surface (Fig. 11.27). The canister walls displayed a similar metal loss and comet-tail-shaped depressions (Fig. 11.2).

Turbulent flow created in the canister and at the canister and tube interface resulted in the metal loss.

---

* Freon is a registered trademark of E. I. du Pont de Nemours & Company.

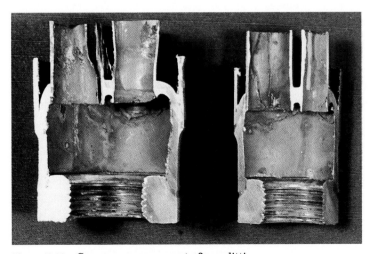

Figure 11.26   Compressor component after splitting.

Figure 11.27   General metal loss surrounding irregular islands of intact metal on the internal surface of the tubes.

CASE HISTORY 11.10

Industry:	Hospital
Specimen Location:	Chilled water control valve
Environment:	Internal: Chilled water
	External: Atmosphere
Time in Service:	Unknown
Sample Specifications:	Cast leaded brass

The cross-sectional schematic of Fig. 11.28 illustrates the spatial relationship between the eroded components shown in Figs. 11.6, 11.29, and 11.30.

Highly localized metal loss at the valve seat is apparent in Fig. 11.6. Figure 11.29 shows the same component close up. Wasted surfaces have a bright, metallic luster free of corrosion products or deposits. Metal loss along the edge of the throttling nut is also apparent (Fig. 11.30).

Erosion-corrosion of these components was caused by high-velocity turbulent flow resulting from incomplete opening of the valve. In this case, erosion is the dominant factor in the metal loss, corrosion being a minor contributing factor.

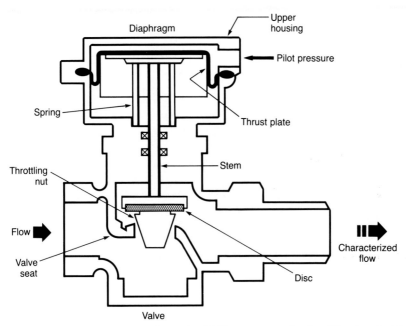

Figure 11.28 Cross-sectional schematic of chill water control valve. (*Reprinted with permission from Landis & Gyr Powers, Inc., Buffalo Grove, Illinois.*)

Figure 11.29 Erosion-corrosion of a valve seat (bright metal in center of photograph).

Figure 11.30   Metal loss by erosion-corrosion at edges of a throttling nut. (Magnification: 7×.)

CASE HISTORY 11.11

Industry:	Steel
Specimen Location:	Bearing-retainer plate from continuous casting roll
Specimen Orientation:	Vertical
Environment:	Treated lake water
Time in Service:	2 years
Sample Specifications:	Carbon steel

Severe, highly localized metal loss on the roll bearing-retainer plate and associated attachment hardware is illustrated in Figs. 11.31 and 11.32. Figure 11.33 illustrates the arrangement of these components in a continuous caster roll system. Note the smooth surface contours at the edge of the plate. Close examination of these surfaces under a low-power stereoscopic microscope revealed fine, wavelike striations.

Exposed ends of the bolts and nuts have also suffered severe localized deterioration, resulting in smooth surface contours. Adjacent areas of each component show no deterioration. Erosion resulting from the impingement of high-velocity slab-cooling spray is the predominant factor in this metal loss.

Figure 11.31   Severe metal loss on the retainer plate at a bolt hole (see Fig. 11.32).

Figure 11.32   Severe localized metal loss on bolt threads and on nuts.

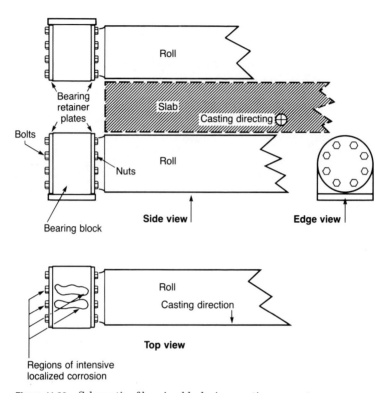

Figure 11.33 Schematic of bearing blocks in a continuous caster.

CASE HISTORY 11.12

Industry:	Chemical process
Specimen Location:	Surface condenser tube
Specimen Orientation:	Horizontal
Environment:	Internal: Cooling water treated with phosphate, dispersant, and bleach; temperature 70–100°F (21–38°C), pH 6.8–7.2
	External: Turbine steam at 850 psi (5.9 MPa), condensate pH 8.0
Time in Service:	15 years
Sample Specifications:	¾ in. (1.9 cm) outside diameter, admiralty brass tube

The tube illustrated in Fig. 11.34 was discovered during a routine inspection of the condenser. The tube had been removed from a position directly in line with the turbine exhaust inlet.

Close examination of each circular spot revealed a small forest of erosion cones (Fig. 11.35). Erosion was caused by the impingement of high-velocity steam probably mixed with droplets of condensate. The circular, equally spaced erosion sites apparently reproduced corresponding, equally spaced circular openings in the exhaust inlet. Compare Case History 11.3.

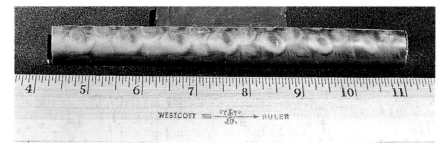

Figure 11.34　A pattern of erosion spots on the external surface facing the turbine exhaust inlet.

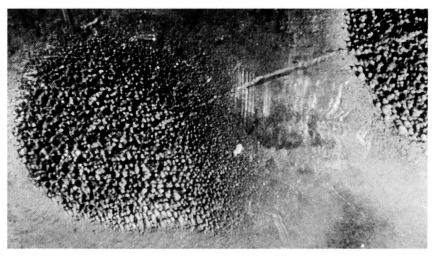

Figure 11.35　A small forest of erosion cones typical of the appearance of each site. (Magnification: 7.5×.)

Cavitation Damage

General Description

The first recorded observation of cavitation was made in 1754 by Euler. However, serious research into the character and mechanism of cavitation did not begin until 1919. Research since that time has established a reasonably sound basis for describing and explaining the phenomenon.

Erosion and cavitation both can degrade materials simply by mechanical means or by combining the effects of mechanical deterioration and corrosion to produce a synergistic result. However, the mechanisms by which erosion and cavitation operate, and the resulting damage, are quite distinct.

Cavitation may be defined as the instantaneous formation and collapse of vapor bubbles in a liquid subject to rapid, intense localized pressure changes. Cavitation damage refers to the deterioration of a material resulting from its exposure to a cavitating fluid.

Figure 12.1 is a simplified representation of the cavitation process. Figure 12.1A represents a vessel containing a liquid. The vessel is closed by an air-tight plunger. When the plunger is withdrawn (B), a partial vacuum is created above the liquid, causing vapor bubbles to form and grow within the liquid. In essence, the liquid boils without a temperature increase. If the plunger is then driven toward the surface of the liquid (C), the pressure in the liquid increases and the bubbles

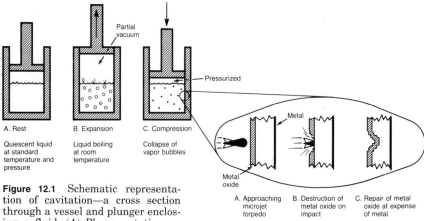

Figure 12.1 Schematic representation of cavitation—a cross section through a vessel and plunger enclosing a fluid. (*A*) Plunger stationary, liquid at standard temperature and pressure; (*B*) plunger withdrawn, liquid boils at room temperature; (*C*) plunger advanced, bubbles collapse.

Figure 12.2 Disintegration of protective corrosion product by impacting microjet torpedo.

condense and collapse (implode). In a cavitating liquid these three steps occur in a matter of milliseconds.

Cavitation damage results from hydrodynamic forces created by the collapsing vapor bubble. Implosion of a vapor bubble creates a microscopic "torpedo" of water that is ejected from the collapsing bubble at velocities that may range from 330 to 1650 ft/s (100–500 m/s). Normally, the energy of this "torpedo" is quickly absorbed by the surrounding fluid. However, if the bubble collapse occurs adjacent to a metal surface, the "torpedo" may impact on the surface, dislodging protective corrosion products (Fig. 12.2) and/or locally deforming the metal itself. Single impacts result in little, if any, metal damage, but continuous impacts can disintegrate protective coatings, thereby forcing the metal to reform the coating continuously by otherwise innocuous corrosion. When the metal is affected, continuous impacts cause ductile or brittle rupture of microscopic chunks of metal. They may induce a highly localized metal fatigue, which also causes eventual, highly localized metal disintegration.

Cavitation damage in the complete absence of corrosion has been demonstrated (e.g., roughening of polished glass in cavitating distilled water). However, in industrial situations it is highly probable that corrosion is a common contributing factor.

Metal damage due to cavitation can be both rapid (Figs. 12.3*A* through *D*) and severe. However, cavitation damage can have a time dependency; see Fig. 12.4. An incubation period may be observed

Figure 12.3A Progressive deformation in aluminum foil exposed to a cavitating fluid for successively longer periods. (*A*) No exposure; (*B*) 5 s; (*C*) 10 s; (*D*) 20 s. (350× SEM.)

during which little or no damage is apparent. Following the incubation period, the rate of metal loss may increase sharply. After reaching this maximum value, the rate of damage may then either decline to a lower steady-state value or fluctuate unpredictably.

The actual shape of the curve illustrated in Fig. 12.4 can vary according to factors such as metal properties and cavitation intensity. (Cavitation intensity relates to the number of bubbles created in a unit volume of fluid and the amount of energy transferred during the col-

Figure 12.3B As in Fig. 12.3A, but exposed for 5 seconds.

Figure 12.3C   As in Fig. 12.3A, but exposed for 10 seconds.

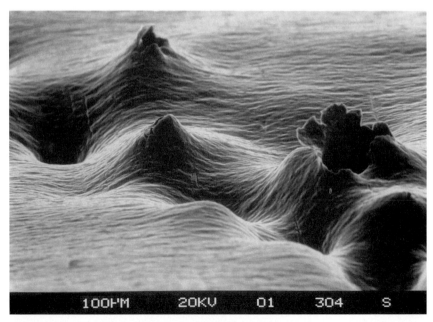

Figure 12.3D   As in Fig. 12.3A, but exposed for 20 seconds.

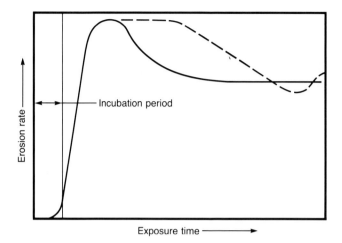

Figure 12.4   Schematic representation of typical variation of cavitation damage rate with exposure time. (*Reprinted with permission of American Society for Metals from* Metals Handbook, *vol. 10, 8th ed., Metals Park, Ohio, 1974, p. 162.*)

lapse of a bubble.) At low levels of cavitation intensity, corrosion reactions may influence the shape of the curve.

Locations

In general, cavitation damage can be anticipated wherever an unstable state of fluid flow exists or where substantial pressure changes are encountered. Susceptible locations include sharp discontinuities on metal surfaces, areas where flow direction is suddenly altered (Fig. 12.5), and regions where the cross-sectional areas of the flow passages are changed.

Damage will be confined to the bubble-collapse region, usually immediately downstream of the low-pressure zone. Components exposed to high velocity or turbulent flow, such as pump impellers and valves, are subject. The suction side of pumps (Case History 12.3) and the discharge side of regulating valves (Fig. 12.6 and Case History 12.4) are frequently affected. Tube ends, tube sheets, and shell outlets in heat exchanger equipment have been affected, as have cylinder liners in diesel engines (Case History 12.1).

Critical Factors

The essential condition governing cavitation is an unstable fluid flow. An unstable fluid flow is, in turn, affected by the properties of the fluid

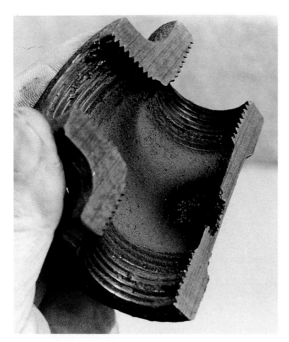

Figure 12.5 Cavitation damage at bottom of tee at impact site of in-flowing water.

Figure 12.6 Cavitation damage on the discharge side of a regulating valve.

itself, the physical characteristics of the surfaces containing the liquid, and equipment operating practices.

Properties of the fluid

Temperature, air content, pressure, and chemical composition of the fluid can affect the tendency of the fluid to cavitate. For example, the presence of minute air bubbles in the fluid can act as nucleation sites for cavitation bubbles, thereby increasing the tendency of the fluid to cavitate. Increasing pressure decreases susceptibility to cavitation; decreasing pressure increases susceptibility to cavitation.

Physical characteristics of container surfaces

Cavitation is favored by:

- Rough rather than smooth surfaces
- Sharp surface discontinuities
- Rapid change in direction
- Abrupt changes in cross section

Equipment operating practices

Circumstances that may induce cavitation include vibration, excessively high flow rates, insufficient supply of fluid to the suction side of pumps, and operation of valves in a partially open or closed position.

Identification

Cavitation typically produces sharp, jagged, spongelike metal loss, even in ductile materials. The affected region is free of deposits and accumulated corrosion products if cavitation has been recent.

Typically, affected surfaces are highly localized to specific regions, although if cavitation is severe and widespread, the area affected may be extensive (Fig. 12.7). On symmetrical components having repeated elements, the pattern of damage may repeat itself at identical locations on each element (Fig. 12.8).

Examination of the microstructure of the cavitated surface will commonly disclose evidence of deformation such as twins (Neumann bands) in carbon steel and general cold working in other metals (Case History 12.6). Damage from cavitation can be differentiated from attack by a strong mineral acid, which can produce a similar surface appearance, by observing the highly specific areas of attack characteristic of cavitation. Acid attack is typically general in its extent (Case History 12.6).

Figure 12.7   Cast iron pump impeller with severe cavitation damage.

Figure 12.8   Cavitation damage repeated on successive elements of a bronze impeller.

Elimination

Several approaches are available to alleviate or eliminate cavitation damage problems:

- Change of materials
- Use of coatings
- Alteration of environment
- Alteration of operating procedures
- Redesign of equipment

Change of materials

Historically, numerous attempts have been made to correlate resistance to cavitation damage with certain material properties such as hardness. Unfortunately, cavitation damage is too complex a phenomenon to permit such simple relationships. Resistance to cavitation damage is significantly influenced by factors such as the corrosive nature of the cavitating fluid and the intensity of the cavitation. Hence, no specific material properties may be used to rank metals according to cavitation resistance in all environments. It is observed, however, that resistance relates in a general way to a metal's hardness and ductility. Consequently, a generalized ranking from most resistant to least resistant is as follows:*

Stellites

304 stainless steel

Nickel-aluminum bronze

Magnesium bronze

Steel

Rubber

Cast iron

Aluminum

Weld overlays of stainless steel or cobalt-based wear-resistant and hard-facing alloys such as Stellite may salvage damaged equipment. In addition, weld overlays incorporated into susceptible zones of new equipment may provide cost-effective resistance to cavitation damage.

---

* From the *Metals Handbook,* vol. 13, 9th ed., 1987, p. 1297.

Use of coatings

Resilient materials such as rubber and some plastics may be useful in certain applications, especially under conditions of low cavitation intensities. However, such materials are subject to disbondment at the metal and elastomer interface at high cavitation intensities, even if the exposure is brief.

Alteration of environment

If the cavitation intensity is low and corrosion is a significant accelerating factor, appropriate inhibitors can be useful. Notable successes have been achieved with diesel engine cylinder liners.

Alteration of operating procedures

In principle, altering operating procedures involves minimizing the frequency and magnitude of localized pressure changes. In practice, this involves such things as eliminating or reducing the magnitude and frequency of vibration in a cylinder liner, maintaining adequate feed to the suction side of a pump impeller (maintaining net positive suction head), or reducing flow velocity through a heat exchanger. (Note that an insufficient supply of fluid to the intake of a pump may be caused by plugged filters upstream of the pump. In this case, alteration of the operating procedure may consist simply of periodically cleaning the filter system.)

If such steps cannot be implemented or are unsuccessful in reducing or eliminating cavitation, injecting air into the cavitating system may be of value. Injected air bubbles can act as an energy-absorbing cushion, thereby minimizing the damage associated with collapsing vapor bubbles.

Redesign of equipment

Of the various approaches to reducing cavitation damage, redesign of equipment is probably the most successful. The following have proven useful:

- Increase radii of curvature
- Increase system pressure
- Remove surface discontinuities
- Blend surface contours to reduce bubble nucleation sites
- Eliminate abrupt changes in cross-sectional areas of tubes, pipes, and so on

Cautions

It is possible to confuse corrosion by a strong mineral acid with damage by cavitation. Both can produce jagged, undercut, spongelike surface contours. The two can be distinguished, however, by noting that attack by a strong acid generally affects all exposed surfaces and is therefore not as location specific. Cavitation is localized to regions experiencing rapid pressure changes and is therefore quite location specific. See Case History 12.6.

Related Problems

See also Chap. 7, "Acid Corrosion."

CASE HISTORY 12.1

Industry:	Automotive
Specimen Location:	Diesel engine cylinder liner
Specimen Orientation:	Vertical
Environment:	Internal: Hot combustion gases, moving piston
	External: Cooling water containing ethylene glycol
Time in Service:	Unknown
Sample Specifications:	Gray cast iron

Figure 12.9 illustrates the typical appearance of cavitation damage along opposite sides of a diesel engine cylinder liner. The vertical pit alignment, the presence of the pits at opposite sides of the liner, and the comparatively smaller population of pits along one side are typical of cavitation damage of cylinder liners. Note also the jagged, spongelike, highly localized metal loss shown in Fig. 12.10. This is characteristic of cavitation damage in general.

Cavitation occurred on the cooling water side due to vibrations generated during the operation of the engine. (Sources of such vibration include piston slap, loose bearings, and a crank shaft or piston that is out of balance.) Experience has indicated that cavitation intensity in these cases is relatively low. Consequently, cavitation destroys the protective iron oxide covering of the metal. The iron oxide spontaneously reforms over the bare metal but is subsequently removed by impacting microjets from cavitation. Hence, the cavitation damage with respect to the metal is indirect, affecting the iron oxide covering rather than the metal itself.

Reduction or elimination of this problem can be effected through reduction or elimination of the vibration. Corrosion inhibitors added to the coolant have also been successful.

Figure 12.9 Typical vertical alignment of cavities resulting from cavitation damage in a diesel engine cylinder liner.

Figure 12.10 Jagged, cavernous pits typical of cavitation damage.

CASE HISTORY 12.2

Industry:	Automotive
Specimen Location:	Diesel engine spacer
Specimen Orientation:	Vertical
Environment:	Cooling water containing ethylene glycol
Time in Service:	200 hours
Sample Specifications:	Cast aluminum

Figure 12.11 illustrates the throat of the spacer. Damage is localized to both sides of a 2¼-in.-long vertical opening.

The ¹⁄₁₆-in.-diameter pits are clustered into groups near the center and along the edges of the bore (Fig. 12.12). The interior of the pits is jagged and free of deposits or visually observable corrosion products.

However, examination of cross-sectional profiles of the pits under a high-power microscope revealed a thin coating of corrosion product and no evidence of cold working of the surface metal. It is probable, therefore, that cavitation affected the coating rather than the metal directly. The coating spontaneously reformed over bare metal surfaces, only to be dislodged by subsequently impacting microjets. In a case such as this, use of an appropriate inhibitor may reduce cavitation damage.

The cavitation damage in this spacer was due to vibrations from operation of the engine. The localized nature of the damage in this case is an illustration of a common feature of cavitation. Pits formed by initial cavitation damage become preferred sites for the development of subsequent cavitation bubble formation due to the jagged, irregular contours of the pit. This tends to localize and intensify the cavitation process, especially in later stages of pit development.

Figure 12.11   Zones of cavitation damage segregated to both sides of a vertical opening.

Figure 12.12   Spongelike surface contour typical of cavitation damage. (Magnification: 8×.)

CASE HISTORY 12.3

Industry:	Pulp and paper
Specimen Location:	Suction side component of cooling tower pump
Specimen Orientation:	Vertical
Environment:	Cooling tower water
Time in Service:	Unknown
Sample Specifications:	Pearlitic gray cast iron

Figure 12.13 illustrates severe damage suffered by a component of a cooling tower water pump. The jagged, undercut, spongelike metal loss characteristic of cavitation damage is apparent in Fig. 12.14. All damage occurred along the inner curvature of the specimen.

Microstructural examinations revealed graphitic corrosion (see Chap. 17) of the metal surfaces. The evidence of graphitic corrosion indicates one of the following:

1. Cavitation is occurring intermittently, with long intervals between occurrences.
2. Cavitation is more or less continuous, but the cavitation intensity is so low that only the coating of corrosion products is dislodged. Resistance of the corrosion products to cavitation relative to the resistance of the metal to cavitation is low, making the corrosion products more susceptible to damage.

Severe cavitation damage on the suction side of the pump reveals insufficient water supply to the pump (insufficient net-positive suction head). Such a circumstance could be caused by partially clogged filters or screens upstream of the pump, or simply by insufficient feed of water to the pump.

Changing the pump metallurgy to a more corrosion- and cavitation-resistant material, such as stainless steel, is a potential solution to this type of problem. Note, however, that all other cast iron pump components that have sustained graphitic corrosion should be replaced to avoid the possibility of galvanic corrosion (see Chap. 16) between retained graphitically corroded cast iron components and new components.

Figure 12.13   Severe damage to a cast iron pump component.

Figure 12.14   Typical appearance of severe cavitation damage in cast iron.

CASE HISTORY 12.4

Industry:	Steel
Specimen Location:	Main supply line to spray cooling system
Specimen Orientation:	Horizontal
Environment:	Internal: Untreated lake water intermittently chlorinated
	External: Air
Time in Service:	6 years
Sample Specifications:	Cast carbon steel

The damage illustrated in Fig. 12.15 occurred in an expansion flange located immediately downstream of a valve. The valve was operated in a partially closed position.

Jagged, spongelike metal loss was present downstream of a weld bead (Fig. 12.16), and again in the region of diameter increase (Fig. 12.17). Metal loss was most severe near the weld bead, where weld repairs from the external surface had closed a large perforation.

The damage was apparently due to severe turbulent flow past a partially closed valve, coupled with flow disruption across the weld bead and pressure reduction in the region of diameter increase.

Potential remedial measures include:

- Operating the valve in the fully open position
- Specifying a cavitation-resistant material such as hardened carbon steel, hardened stainless steel, or carbon steel overlaid with a cavitation-resistant material

Figure 12.15   Damage to the internal surface of a supply line.

Figure 12.16   Jagged metal loss downstream of protruding weld bead.

Figure 12.17   Metal loss downstream of flange wall expansion.

CASE HISTORY 12.5

Industry:	Air conditioning
Specimen Location:	Chiller condenser tube
Specimen Orientation:	Horizontal
Environment:	Internal: Cooling water, pH 8.6, all organic treatment
	External: Freon refrigerant
Time in Service:	Unknown
Sample Specifications:	¾ in. (1.9 cm) outside diameter externally finned, copper tubes

Pinhole perforations were discovered in the walls of chiller condenser tubes of an air-conditioning system. Close laboratory examination of the internal surfaces of affected tubes revealed distinct patches of small pits (Fig. 12.18) and pit sites aligned along longitudinal mandrel marks and fine scratches (Fig. 12.19). In some locations, transversely oriented pit sites that were aligned with the locations of the fins on the external surface branched off the primary longitudinal pit alignment (Fig. 12.20).

Examination of surface profiles in these pitted regions under a high-power microscope revealed jagged, undercut profiles free of deposits or corrosion products. This appearance is typical of cavitation damage.

This case vividly illustrates the potential effect of surface roughness on the propagation of cavitation bubbles. Figure 12.18 represents a surface previously roughened by underdeposit corrosion (see Chap. 4, "Underdeposit Corrosion") that subsequently became a favored site for cavitation because of its roughness. Roughness resulting from mandrel marks and scratches induced the pits of Fig. 12.19. The vertical branching from the primary longitudinal pit alignment in Fig. 12.20 resulted from abrupt contour changes on the internal surface caused by the finning operation performed on the external surface during manufacture of the tubes.

Note that localized corrosion having the appearance illustrated in Figs. 12.18 through 12.20 could be associated with brief exposure to a strong acid. In this case, however, all available information indicated that the tubes had never been exposed to an acid of any type. Cavitation was caused by high-frequency vibration of the tubes. The vibration apparently induced a threshold cavitation intensity such that rough or irregular surfaces produced cavitation bubbles, and smooth internal surfaces did not.

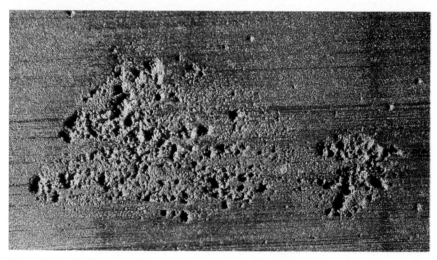

Figure 12.18   Confined patches of small pits. (Magnification: 15×.)

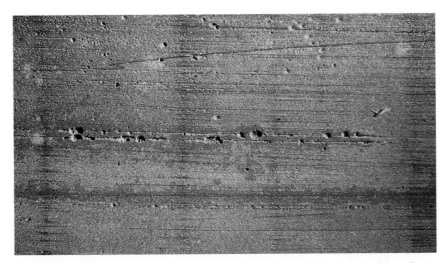

Figure 12.19 Attack sites aligned along mandrel marks or scratches on the internal surface. (Magnification: 7.5×.)

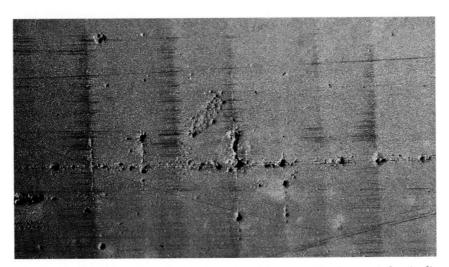

Figure 12.20 Vertical pit sites aligned with external fins branching from main longitudinal trunk. (Magnification: 7.5×.)

CASE HISTORY 12.6

Industry:	Chemical process—gas plant
Specimen Location:	Fin fan condenser
Specimen Orientation:	Horizontal
Environment:	Internal: Steam and condensate at 40 psi (276 kPa), pH 8.5–8.8 with excursions to 6.9
Time in Service:	2½ years
Sample Specifications:	1 in. (2.5 cm) outside diameter, mild steel tube

Recent failures of the type illustrated in Fig. 12.21 affected a total of eight tubes in this condenser. Metal loss occurred exclusively on the top and bottom internal surfaces. Affected areas have a rough, jagged contour of deep, overlapping pits that were essentially free of corrosion products. Unaffected areas of the internal surface are smooth and are covered with a layer of black iron oxide.

Microstructural examinations revealed deformation twins (Neumann bands) in metal grains at wasted surfaces. The surfaces in these areas have a jagged, undercut profile.

The overall appearance of this sample is typical of cavitation damage, although the attack could conceivably be confused with corrosion by a strong, mineral acid. However, at least two distinct features confirm the diagnosis of cavitation damage. First, the attack occurred in opposite areas (top and bottom) of the internal surface. The sides of the tube remain completely unaffected. This is typical of cavitation damage resulting from cyclic vibration of the tube in the vertical plane. Acid attack would affect the entire internal surface.

Second, deformation twins were observed in metal grains at the damaged surfaces. Deformation twinning cannot result from corrosion but is the consequence of shock loading of the metal, precisely the effects of microjets of water impacting on the metal surface.

The longitudinal crack apparent in Fig. 12.21 resulted from stresses from internal pressure that exceeded the tensile strength of the metal in the greatly thinned tube wall.

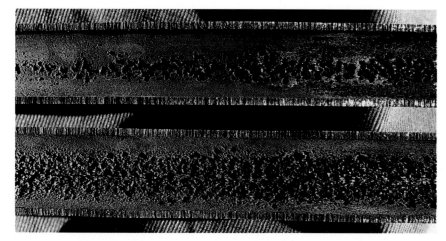

Figure 12.21 Cavitation damage on the internal surface of the condenser tube. Note longitudinal crack. The surfaces are covered with orange, air-formed iron oxides that formed subsequently to the removal of the condenser tube.

13

Dealloying

General Description

Dealloying occurs when one or more alloy components are preferentially removed from the metal. This depletion phenomenon differentiates dealloying from other forms of corrosion. Dealloying is also sometimes referred to as *selective leaching* or *parting*. The specific terminology used comes from the alloying element removed. For example, the preferential leaching of zinc from brass is called *dezincification*. If nickel is removed, the process is called *denickelification,* and so forth.

Dealloyed areas are structurally altered. Corroded areas are weak and porous, causing fracture and weeping leaks.

Dezincification

Admiralty brass (70% Cu, 29% Zn, 1% Sn, 0.05% As or Sb) and arsenical aluminum brass (76% Cu, 22% Zn, 2% Al, 0.05% As) are resistant to dezincification in most cooling water environments. In the recent past, heat exchangers have virtually always been tubed with inhibited grades of brass. Brasses containing 15% or less zinc are almost immune to dezincification. Dezincification is common in uninhibited brasses containing more than 20% zinc. Inhibiting elements include arsenic, antimony, and phosphorus. Without inhibiting elements, admiralty and aluminum brasses are subject to attack. Brasses containing more than about 32% zinc are quite susceptible to attack.

Dezincification is explained by two theories. The first is that the alloy dissolves, with a preferential redeposition of copper. The other proposes a selective leaching of the zinc, leaving the copper behind. There is evidence that both mechanisms may operate, depending on the specific environment.

Denickelification

Selective removal of nickel from copper alloys is common. However, denickelification does not commonly cause the affected component to fail. Rather, the liberated nickel may deposit downstream and/or be released into the environment.

Locations

As the word *dealloying* implies, attack occurs only in metals containing two or more alloying elements. Various alloys are susceptible to corrosion. Copper alloys such as brasses, cupronickels, and bronzes are particularly susceptible in cooling water environments. Most other alloys, with the exception of gray and nodular cast iron (see Chap. 17, "Graphitic Corrosion"), are attacked when exposed to high temperatures, molten salts, acids, sulfides, or other very aggressive environments.

Because of excellent thermal conductivities, suitable mechanical properties, and superior corrosion resistance, brasses, cupronickels, and to a lesser extent bronzes are often used in heat exchangers. Thus, heat exchangers and condensers experience attack. Piping and plumbing fittings, especially threaded unions carrying both hot and cold water, as well as strainers, screens, and gasketed components are frequently affected. Copper-alloy castings containing high concentrations of zinc are susceptible to attack.

Many copper alloys are also used in pumps as bushings, bearings, impellers, and gaskets. Bronzes, brasses, and other copper alloys are frequently used. Thus, pump components are often corroded.

Critical Factors

Dealloying is influenced by many factors. In general, any process that increases general corrosion will promote dealloying. However, specific acceleration factors may be further classified into one of three categories: metallurgy, environment, and water chemistry.

Metallurgy

Zinc brasses containing phosphorus, antimony, and/or arsenic dezinc-ify much less readily than brasses free of these elements. Brasses containing less than 15% zinc are virtually immune to attack, and those containing more than 32% zinc are readily dealloyed.

Denickelification generally produces less wastage in cupronickels than dezincification in brasses. Wastage decreases as nickel content increases, becoming very slight in alloys containing 30% or more nickel.

Environment

Stagnant conditions, deposits, heat transfer, high temperature, crevice conditions, and stress accelerate dealloying. Porous or granular deposits further attack. Surface shielding produced by deposition and crevices promotes differential concentration cell corrosion and allows aggressive chemical concentration. High metal temperatures accelerate chemical corrosion processes, and thermal gradients directly increase local metal dissolution (in the case of many copper alloys). Stagnant conditions usually worsen all the above. Thus, frequent or prolonged outages contribute to severe attack.

Water chemistry

Soft waters, high concentrations of dissolved carbon dioxide, acidic or high pH conditions, and high chloride concentrations promote dezinci-fication. Calcium-carbonate precipitation in many instances reduces corrosion on yellow metals, including dealloying. Acid waters generally accelerate corrosion on a variety of alloys. Very high pH can also accelerate corrosion and thus promote attack. High chloride concentration has a particularly deleterious effect on brasses and cupronickels (though less so on cupronickels). Hence, brasses and cupronickel components contacting brackish waters or sea water are at special risk. Highly stressed components are more susceptible to all forms of attack.

Identification

Dezincification

The two major forms of dezincification are "layer" and "plug." The names are taken from characteristic corrosion-product morphologies. In *layer-type dezincification,* the entire component surface is converted to corrosion product to a roughly uniform depth (Figs. 13.1 through 13.3). *Plug-type dezincification* produces small pockets or plugs of almost pure copper (Figs. 13.4 and 13.5). Surfaces may have to be cleaned of deposits

Figure 13.1 Layer-type dezincification on a brass casting. The red layers are uniformly corroded regions. The original yellow of the brass is visible in between.

to reveal damage, as dealloying frequently occurs beneath such material (see the section "Cautions," later in this chapter).

Layer-type dezincification is easy to recognize visually. The original component shape and dimensions are usually preserved, but the metal color changes from the golden yellow of zinc brass to the red of ele-

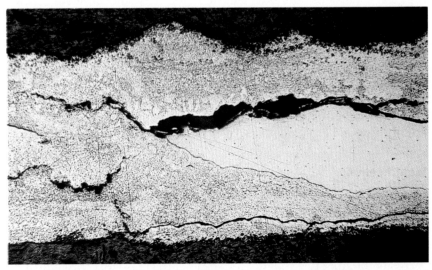

Figure 13.2 Layer-type dezincification of a thin brass plate. The 0.019-in. (0.048-cm) plate is shown in cross section. The dezincified layers converge toward the plate edge. Note the porosity of the dezincified metal.

Figure 13.3 Layer-type dezincification on the internal surface of a cast brass pipe 90° elbow.

mental copper (Figs. 13.6*A* and *B*). The corroded metal has little mechanical strength, and if stressed it fractures easily.

Plug-type dezincification is common on horizontal pipe sections. Attack is usually confined to areas beneath deposits lying along tube bottoms (Figs. 13.4 and 13.7). Plugs are sometimes blown out of tube walls due to internal pressure; holes are produced. More often, fluids weep through the porous plugs, leaving stains or mounds of dissolved

Figure 13.4   A large plug of dezincified metal beneath a deposit. Plugs often form on the bottom of horizontal pipes, beneath deposits.

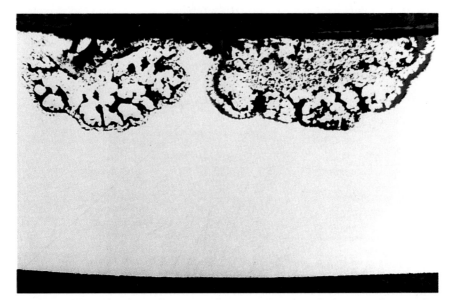

Figure 13.5   Plug-type dezincification on the internal surface of a brass condenser tube. Note the extreme porosity of the copper plugs. Tube wall thickness was 0.040 in. (0.10 cm). Compare to Fig. 13.13. (*Courtesy of National Association of Corrosion Engineers, Corrosion '89 Paper No. 197 by H. M. Herro.*)

solids concentrated by evaporation on external surfaces (Fig. 13.8). Unless internal pressures are high, such plugs may remain in place for long periods with no significant water loss. Close visual inspection of both plug- and layer-type corrosion products reveals stratification of corroded metal (Figs. 13.7 and 13.9). These layers probably represent changes in temperature, water chemistry, and/or corrosion processes.

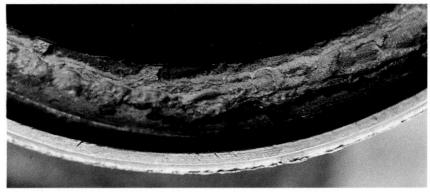

Figure 13.6*A*   Red, dezincified gasketed surface of a cast valve throat. Note the partial spalling of the copper corrosion product.

Figure 13.6*B*   Cross section through the valve throat in Fig. 13.6*A*. Note how the dezincification is confined to the gasket region.

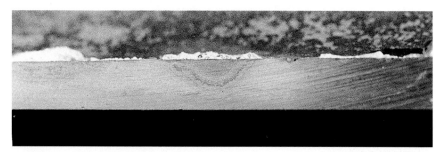

Figure 13.7   Brass tube wall in longitudinal cross section. Plugs of dezincified metal are present on the bottom of the horizontally oriented tube, beneath deposits. Stratification within the copper plugs is evident.

Figure 13.8   A brass tube showing plug-type dezincification. The white patch above the plug on the external surface was caused by dissolved solids, concentrated by evaporation of water leaking through the porous dezincified plug.

Figure 13.9   Stratified copper corrosion product in plug-type dezincification.

Denickelification

Because wastage is usually slight, identification by visual observation alone is difficult; microscopic examination is usually required. Layer-type dealloying commonly occurs. Plug-type denickelification attack has never been observed at this laboratory. Surfaces have a reddish color due to the accumulation of denickelified metal.

Elimination

Dealloying can be reduced (as can any other form of corrosion) by good system operation and the judicious use of appropriate materials and chemical treatment. Specific categories needing attention follow.

Materials substitution

In any specific environment, only certain alloys are affected. Substitution of more resistant materials does not always necessitate major alloy compositional changes. Adding as little as a few hundredths of a percent of arsenic, for example, can markedly reduce dezincification in cartridge brass. Antimony and phosphorus additions up to 0.1% are similarly efficacious.

Surface cleanliness

The propensity toward dealloying decreases as surface cleanliness increases. Flow should be maintained at rates high enough to prevent settling of particulates and to reduce biological attachment and growth (in some cases). Also, chemical treatment effectiveness increases with flow. Surfaces on which fouling is apparent should be regularly cleaned. The use of abrasive sponge balls in condensers has sometimes been effective in keeping tubes clean.

Chemical treatment

Chemical corrosion inhibition can reduce all forms of corrosion including dealloying. In particular, filmers such as tolyltriazole are effective in reducing corrosion of yellow metals.

Biological control

Any deposit can increase dealloying. Biological accumulations, such as slime layers, are no exception to this rule. In the age of increasing public concern regarding pollution, biological control via chemical treatment can be difficult. However, good biological control is always beneficial.

Cautions

Identifying dealloying involves recognizing the variation of alloy composition caused by corrosion. Redeposition of alloy elements can occur during acid cleaning, misleading the examiner. Frequently, copper is redeposited on brasses, bronzes, and similar copper alloys during acid cleaning of deposits.

Related Problems

See also Chap. 2, "Crevice Corrosion"; Chap. 4, "Underdeposit Corrosion"; and Chap. 17, "Graphitic Corrosion."

CASE HISTORY 13.1

Industry:	Mining
Specimen Location:	Turbine condenser
Specimen Orientation:	Horizontal
Environment:	Shell side: Condensing steam
	Tube side: Cooling water, 60–80°F (16–27°C), high hardness and silica, treated with a dispersant and zinc
Time in Service:	6 years
Sample Specifications:	¾ in. (1.9 cm) outer diameter, CDA 442 uninhibited brass tubing

A brass condenser tube contained a ¹⁄₁₆-in. (0.16-cm) diameter hole on the apparent bottom side (Fig. 13.10A). Many other shallow depressions pockmarked internal surfaces. Most depressions were filled with red, copper-colored corrosion products.

The tube was transversely sectioned, polished, and viewed with various microscopes. Large porous corrosion-product plugs extended entirely through the 0.050-in. (0.13-cm)-thick wall (Fig. 13.10B). The depressions contained 94% copper, 5% zinc, and 1% tin.

The depressions and plugs of corroded metal were caused by dezincification. The hole shown in Fig. 13.10A was caused when one such plug blew out of the wall due to the pressure difference between internal and external surfaces.

The condenser was mechanically cleaned every 6 months. Heavy deposition due to silt, sand, and mud plagued this system.

Figure 13.10A   A hole emanating from the internal surface of a brass condenser tube. The hole was caused by plug-type dezincification (see Fig. 13.11).

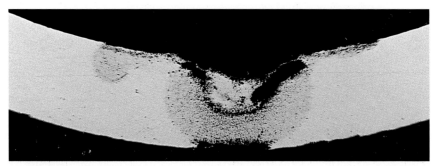

Figure 13.10B   A transverse cross section through the tube wall in Fig. 13.10A. Note the through-wall plug-type dezincification.

CASE HISTORY 13.2

Industry:	Refinery
Specimen Location:	Overhead condenser
Specimen Orientation:	Horizontal
Environment:	Shell Side: Atmospheric pressure, pH 5.5–6.5 (ammonia used to control pH), 240–110°F (116–43°C), overhead vapors
	Tube side: Pressure 30 psi (207 kPa), pH 8.0, 87–120°F (30–49°C), cooling water (no treatment)
Time in Service:	1 year
Sample Specifications:	¾ in. (1.9 cm) outer diameter, uninhibited brass, 72% copper, 26.7% zinc, 1.3% tin

Severe wastage was evident on both external and internal surfaces of the tube section. There were two round holes on one side. Deep pockets of internal surface corrosion penetrated to the external surface (Fig. 13.11A). The depressions contained sulfides, mainly concentrated near external surfaces. Below were large amounts of porous elemental copper (Fig. 13.11B).

Internal surfaces were covered with a tan deposit layer up to 0.033 in. (0.084 cm) thick. The deposits were analyzed by energy-dispersive spectroscopy and were found to contain 24% calcium, 17% silicon, 16% zinc, 11% phosphorus, 7% magnesium, 2% each sodium, iron, and sulfur, 1% manganese, and 18% carbonate by weight. The porous corrosion product shown in Fig. 13.11B contained 93% copper, 3% zinc, 3% tin, and 1% iron. Traces of sulfur and aluminum were also found. Near external surfaces, up to 27% of the corrosion product was sulfur.

The holes and depressions on external surfaces were caused by deep dezincification on internal surfaces. The porous corrosion product plugs had enough structural integrity so that only a few holes formed due to internal pressurization.

The dezincification was caused by underdeposit corrosion. The fact that the brass was not an inhibited grade was a major contributing factor. Chemical cleaning had not been done since this exchanger was installed. No chemical treatment was used on either external or internal surfaces.

Figure 13.11A   Deep, hemispherical "depressions" on the external surface of an overhead condenser due to severe dezincification from internal surfaces (see Fig. 13.11B).

Figure 13.11B   Porous copper corrosion products filling an internal surface depression. The edge of one crater seen in Fig. 13.11A is to the right.

CASE HISTORY **13.3**

Industry:	Utility
Specimen Location:	Main condenser
Specimen Orientation:	Horizontal
Environment:	Tube side: 60–80°F (16–27°C), pH 8.3–8.7; treatment—tolyltriazole, dispersant; brackish water
Time in Service:	3 years
Sample Specifications:	⅞ in. (2.2 cm) outer diameter, 90:10 cupronickel, 0.031 in. (0.079 cm) wall thickness

Several sections of 90:10 cupronickel condenser tubing were received. The sections were from tubes that had been plugged because of earlier failures. No tube section contained the original tube failure, however.

Internal surfaces were covered by loosely adherent corrosion product and deposit. Much of the corrosion product was cuprous oxide. Substantial amounts of iron, silicon, aluminum, zinc, and nickel were also found. Not unexpectedly, chlorine concentrations up to 2% by weight were present; sulfur concentrations of about 1% were also found.

Removal of deposits and corrosion products from internal surfaces revealed irregular metal loss. Additionally, surfaces in wasted areas showed patches of elemental copper (later confirmed by energy-dispersive spectroscopy) (Fig. 13.12). These denickelified areas were confined to regions showing metal loss. Microscopic analysis confirmed that dealloying, not just redeposition of copper onto the cupronickel from the acid bath used during deposit removal, had occurred.

Surfaces of plugged tubes were carefully compared to surfaces of tubes from the same condenser that were never plugged. Only plugged tubes showed denickelification. Stagnant conditions, as well as the deposits and chloride, caused the dealloying. The deepest dealloying was up to 5% of the remaining wall thickness.

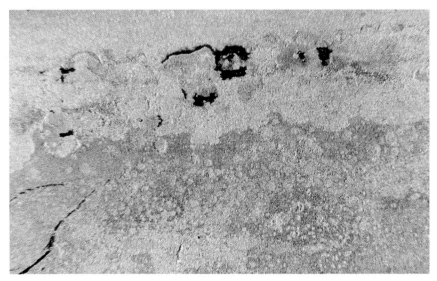

Figure 13.12 Close-up of a 90:10 cupronickel condenser's internal surface after acid cleaning. Elemental copper stains mark regions of shallow denickelification.

CASE HISTORY 13.4

Industry:	Utility
Specimen Location:	Condenser tubes
Specimen Orientation:	Horizontal
Environment:	Tube side: 62–80°F (17–27°C), pH 8.3–8.6; treatment—tolyltriazole, dispersant
Time in Service:	1 year and 12 years (two different samples)
Sample Specifications:	ASTMB395 (71.4% copper, 0.97% tin, <0.005% phosphorus, <0.01% arsenic,others total <0.15%, balance zinc)

Two sections of utility condenser tubing were received. One of the sections had deep plug-type dezincification on internal surfaces (Fig. 13.5); the other showed only superficial corrosion on internal surfaces (Fig. 13.13).

Sections were removed from two similar condensers. Both condensers were fed with a common cooling water source, were of identical design, and experienced virtually identical operating conditions. However, the first exchanger had been in service for 12 years, whereas the other was only on line for 1 year. The dezincified tube (Fig. 13.5) was from the exchanger seeing short-term service. The dezincified tube was fabricated from an uninhibited grade of brass. The older, superficially corroded section contained 0.021% arsenic, an inhibiting agent.

This case history dramatically illustrates the value of proper alloy choice. It was found that the failed exchanger had been ordered many years ago but was not installed until recently. Today, uninhibited grades of brass are almost never used for condenser and heat exchanger service in the United States.

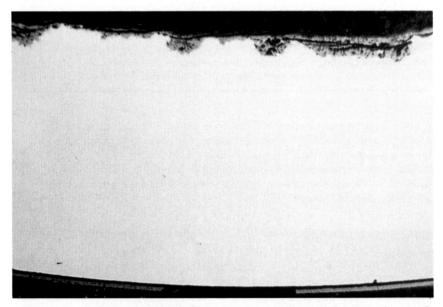

Figure 13.13   An inhibited-brass condenser tube in cross section, showing only superficial wastage on internal surfaces. This tube was exposed to virtually identical conditions as the tube shown in Fig. 13.5. However, because of the inhibition, dezincification did not occur on this tube. (*Courtesy of National Association of Corrosion Engineers, Corrosion '89 Paper No. 197 by H. M. Herro.*)

CASE HISTORY 13.5

Industry:	Hotel
Specimen Location:	Pump impeller and shaft bushing
Specimen Orientation:	Vertical impeller, horizontal bushing
Environment:	Cooling water (brine well), chloride 22,000 ppm, sulfate 2400 ppm, calcium 1700 ppm, magnesium 4300 ppm, iron 0.05 ppm, silica 14 ppm, pH 7.4, 70°F (21°C), pressure 60–70 psi (0.4–0.5 MPa)
Time in Service:	~1 year
Sample Specifications:	6½ in. (16.5 cm) diameter, bronze impeller, 1 in. (2.5 cm) diameter, leaded brass bushing

A pump impeller and a shaft bushing from a small cooling water pump assembly were generally corroded. Reddish surface discoloration revealed layer-type dezincification (Figs. 13.14 and 13.15).

Attack occurred because of substantial idle periods in which flow was limited. Also, very high concentrations of chloride and sulfate in cooling waters accelerated attack.

Figure 13.14 A generally dezincified bronze pump impeller. Surfaces are dealloyed to a depth of about 0.020 in. (0.050 cm).

Figure 13.15   The tricolor internal surface of a leaded-brass bushing. Greenish-blue corrosion product overlies the red of the generally dezincified surface. To the extreme right, only bare metal is visible due to erosion.

Introduction to Material and Weld Defects

The true influence of flaws and defects on component failure is commonly misunderstood. This misunderstanding often arises from one of two misconceptions. The first misconception can be clarified by simple definitions of a flaw and a defect.

Flaw. *A* flaw *can be defined as an imperfection in a material that does not affect its usefulness or serviceability. A component may have imperfections and still retain its usefulness. This fact is recognized by most material codes that permit, but limit, the size and extent of imperfections. This is particularly true of welds, which commonly contain harmless imperfections. It is not uncommon for failures to occur in the vicinity of flaws that have contributed nothing to the failure mode.*

Defect. *A* defect *is an imperfection in a component, resulting from the way it was manufactured, shipped, handled, or installed, that may contribute materially to failure or limited serviceability. Failures resulting directly or indirectly from defects account for less than 1% of all failures.*

The second misconception involves the perception of what constitutes a defect. A defect is not simply a visually observable discontinuity such as a hole, lap, or seam in a component. Defects, from a failure-analysis standpoint, may also be such things as a high residual stress that may lead to cracking or unfavorable microstructures that can lead to either cracking or highly localized corrosion. These considerations will be covered in Chap. 14, "Material Defects," and Chap. 15, "Weld Defects."

Chapter 14

Material Defects

General Description

The percentage of failures that occur solely from material defects is quite low—less than 1% of all failures. What is more common is that defects may act in conjunction with specific environmental factors to produce failures, such as cracking or localized corrosion.

Nearly every manufacturing, shipping, and installation process is a potential source of a defect. Manufacturers, in recognition of this, mount major efforts to minimize or eliminate defects. Although an exhaustive list of defects is beyond the scope of this book, some common ones will be discussed.

Defects in finished components can sometimes be traced as far back as casting problems in the ingot or billet from which the component was eventually produced. Defects in welded tubes may be linked to defective metal or to the welding process itself. Defects in seamless tubes can originate in extrusion or drawing processes.

Defects in tubes can appear on the surfaces or be hidden within the tube wall. Some common defects are listed here and described in the section on identification.

1. *Gouges.* Gouges are usually elongated grooves in the tube wall caused by mechanical removal of metal during tube fabrication.

2. *Laminations.* A lamination is a separation within the metal wall caused by the presence of a highly elongated nonmetallic inclusion. Laminations generally are aligned parallel to the surfaces.

3. *Laps.* Laps are folds of metal that have been rolled into the metal surface but are not fused to the metal.

4. *Seams.* As defects, seams are distinct from weld seams and may be found in nonwelded (seamless) tubes. They are caused by crevices that have been closed by some rolling process but remain unfused.

5. *Weld-seam defects.* Failure to fuse the metal fully along the weld line in welded tubes may result in a linear open seam or crevice.

Surface defects, if sufficiently severe, may result in failure by themselves. More commonly, they act as triggering mechanisms for other failure modes. For example, open laps or seams may lead to crevice corrosion or to concentration sites for ions that may induce stress-corrosion cracking.

Grooves, laps, and burrs from fabrication processes may act as stress-concentration sites leading to fatigue, corrosion fatigue, or stress-corrosion cracking. Some manufacturing activities such as drawing, straightening, heat treating, welding, bending, flaring, and even rough handling can produce invisible defects, such as harmful levels of residual stress, that can lead to failure by any of several cracking modes.

Dents in tubing can induce erosion failures, especially in soft metals such as copper and brass. Welding and improper heat treatment of stainless steel can lead to localized corrosion or cracking through a change in the microstructure, such as sensitization. Another form of defect is the inadvertent substitution of an improper material.

Locations

In general, manufacturing defects can exist in any component. However, defects resulting from fabrication or installation activities tend to occur in specific locations, such as:

- Tube ends where they are rolled into tube sheets
- Bends, especially the innermost bend of U-tube bundles
- Welds
- Seams in welded tubes

Critical Factors

The critical factors governing defect-related failures are the severity of
the defect and the environmental conditions with which the defect may
interact to produce failure. A defect, if sufficiently severe, may cause
failure in and of itself, without contributions from environmental conditions. Defect-environment interactions are more common, however.
Unquestionably, equipment can operate satisfactorily with components containing significant defects simply because there are no environmental conditions with which the defect may interact to produce
failure.

This explains why equipment may operate satisfactorily for years,
then fail shortly after some seemingly innocuous change is made in
the operating environment. For example, a dent in a brass heat
exchanger tube may remain harmless at coolant flow velocities of
3 ft/s (0.9 m/s) but may become the site of erosion at 5 ft/s (1.5 m/s).
Residual stress in a brass tube at a bend or at the rolled end of the
tube may provide years of trouble-free service until the environment
is contaminated with ammonia and failure occurs by stress-corrosion
cracking shortly thereafter. Laps, seams, or sensitized metals in a
stainless steel component may remain undetected until a change in
operation introduces chloride ions, after which pitting, cracking, or
localized corrosion may produce a failure. Hence, defect-related failures tend to be opportunistic.

Identification

Most defects can be detected using one or more appropriate nondestructive testing techniques. However, in the absence of routine nondestructive testing inspections, identification of defects in installed
equipment is generally limited to those that can be observed visually.
Defects such as high residual stresses, microstructural defects such as
sensitized welds in stainless steel, and laminations will normally
remain undetected. Defects that can be detected visually have the following features:

1. *Gouges.* Gouges are often elongated grooves in a tube wall.
   Depending on the cause of the gouge, the groove may be very shallow or relatively deep and may be rounded or sharp. Sharp grooves
   can act as stress concentrators, which can lead to various kinds of
   cracking. Gouges can be caused by mechanical removal of metal during tube fabrication (Case History 14.4).

2. *Seams.*   As a defect, a seam is distinct from the seam resulting from a welding process. Seam defects can be found in nonwelded (seamless) tubes. They can originate from blow holes or nonmetallic inclusions in the ingot and are caused by crevices that have been closed by some rolling process but remain unfused. At times, they will appear in a spiral pattern in tubes. Seams can be very tight and appear as hairlines on the surface. They can cause failure when the component is pressurized.

3. *Laps.*   Laps are folds of metal that have been rolled into the metal surface but are not fused.

4. *Dents, crimps, and flattening.*   These appear as simple deformation of the normal geometry of the component.

5. *Weld-seam defect.*   This appears as a linear groove or crevice running along the seam formed in a welded component.

Elimination

Elimination of defect-related failures has a limited chance of success once equipment is installed. The key to successful elimination is to prevent defective components from being installed in the first place. This is accomplished in one of two ways:

1. Acquire equipment produced under successful zero defect programs.

2. Adopt and follow stringent quality control practices.

Stringent quality control practices require adequate inspections at all stages, including manufacturing, shipping, handling, and installation. In addition to visual inspections, several nondestructive testing techniques are available to facilitate inspections. However, it should be recognized that no individual or series of nondestructive testing techniques is capable of detecting each and every defect.

Among the available nondestructive testing techniques are ultrasonic tests, eddy-current tests, air-underwater and hydrostatic tests, and visual examinations. All of these techniques must be executed by technically qualified personnel.

Each test has specific limitations and may pass defects of various sizes and orientations. The specific limitations for a particular technique must be understood for correct interpretation of results. For example, hydrostatic tests and air-underwater tests may pass defects that run 70% of the wall thickness. Hydrostatic tests may pressurize a defect in such a way that leaking does not occur in the test but can in service.

Water used in any testing procedure should be free of corrodents and of substances whose decomposition products may be corrosive. Metal surfaces should be dried after testing and kept dry prior to service.

Cautions

Distinctions need to be drawn between simple material flaws and service-threatening defects. Rejection of components that are merely flawed can be economically unjustifiable. Distinguishing between flaws and defects generally requires the judgment of an experienced inspector.

Related Problems

See also Chap. 15, "Weld Defects."

CASE HISTORY 14.1

Industry:	Chemical process
Specimen Location:	Heat exchanger
Specimen Orientation:	Horizontal
Environment:	Internal: Cooling water: phosphate, zinc, and polymer treatment, 70°F (21°C)
	External: Gaseous environment containing ammonia, 250°F (121°C)
Time in Service:	10 years
Sample Specifications:	¾ in. (1.9 cm) outside diameter, carbon steel

Figure 14.1 illustrates one of many similar pinhole perforations found in this and other tubes within the exchanger. Figure 14.2 shows the typical appearance of the internal surface immediately after removing the tube from the exchanger. Laboratory acid cleaning of the internal surface revealed a defective weld seam (Fig. 14.3).

All of the perforations occurred at sites of small pits that were aligned along the seam. All perforations occurred within a few feet of the tube ends, although pits were present within the seam along the entire length of the tube.

Microstructural examinations revealed V-shaped openings along the tube seam, some extending into as much as 50% of the tube wall thickness. The incompletely closed seam provided a crevice in which differential concentration cells developed (see Chap. 2, "Crevice Corrosion"). The resulting localized corrosion caused the observed pits.

Longitudinal alignment of perforations in a welded tube is a strong indication that a seam may be defective.

Figure 14.1 Pinhole perforation in a steel heat exchanger tube, viewed from the external surface.

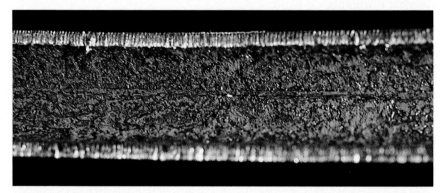

Figure 14.2 Typical appearance of the internal surface. Note the small pit centered on the longitudinal seam.

Figure 14.3 Close-up of the internal surface following acid cleaning. Note the perforation in the tube wall along the weld seam.

CASE HISTORY 14.2

Industry:	Chemical process
Specimen Location:	Cooling water bypass line
Environment:	Internal: Cooling water: all-organic treatment, 95°F (35°C), pH 8.5–9.2, M alkalinity 700–800 ppm
	External: Ambient
Time in Service:	Unknown
Sample Specifications:	2⅜ in. (6 cm) outer diameter, carbon steel

The tubercle apparent on the internal surface in Fig. 14.4 is located directly above a deep, longitudinal groove that runs the length of the internal surface. The tubercle caps a perforation in the tube wall. The tubercle is soft and easily dislodged. Other, smaller tubercles are aligned along the seam.

Chemical removal of deposits and corrosion products revealed the appearance of the groove (Fig. 14.5). The crevice formed by the incompletely fused weld seam fostered the establishment of differential concentration cells (see Chap. 2). This resulted in localized corrosion and eventual perforation through the greatly thinned tube wall at the bottom of the crevice. The tubercle, which is composed of corrosion products, is a simple result of the corrosion process occurring locally within the crevice.

Figure 14.4   Tubercle capping a pit located in a weld-seam defect.

Figure 14.5   Severe seam defect apparent after acid cleaning of the internal surface.

CASE HISTORY 14.3

Industry:	Air conditioning
Specimen Location:	Evaporator and chiller tube
Specimen Orientation:	Horizontal
Environment:	Internal: Treated cooling water: 50°F (10°C), pressure 45 psi (310 kPa)
	External: Freon refrigerant 44°F (7°C)
Time in Service:	10 years
Sample Specifications:	¾ in. (1.9 cm) outer diameter, externally finned copper tube

Figure 14.6 shows the straight, unbranched, longitudinal cracks typical of tubes removed from the unit. The cracked tubes had been located by eddy-current inspection.

Close examination revealed that the cracks originate on the external surface. Exposure of the fracture surface revealed a rough contour covered with dark copper oxide. Close examination of the internal surface revealed fewer, tighter fissures corresponding to the locations of cracks on the external surface.

No other evidence of deterioration, such as corrosion, was apparent on either surface. The cracks are probably material defects. They may be laps or seams that were present on the external surface prior to the fin-rolling operation and were exaggerated during the rolling process.

Figure 14.6 Material defect, perhaps a lap or seam, on the external surface of a finned copper tube.

CASE HISTORY 14.4

Industry:	Utility
Specimen Location:	Condenser tube
Environment:	Ambient conditions, new tube
Time in Service:	New tubes—several months in storage
Sample Specifications:	1 in. (2.5 cm) outside diameter, 90:10 cupronickel seamless tube

The various forms of defects on the internal surface of this tube could be overlooked in a casual, visual examination. Closer observation, however, would disclose several forms of discontinuities, such as shallow gouges (Fig. 14.7) and particles of smeared metal (Fig. 14.8). These features are prominent in a distinct longitudinal zone (Fig. 14.9).

Such defects result from abnormal manufacturing operations such as insufficient lubrication between the metal and the mandrel during the tube-forming process. The lubricant may have been contaminated. Measurement indicated that some of these defects penetrated 8% of the tube wall thickness. Defects of this type can act as corrosion-initiation sites in a sufficiently aggressive environment.

Figure 14.7   Shallow, aligned gouges on the internal surface. (Magnification: 6.5×.)

Figure 14.8   Broad patches of smeared metal aligned longitudinally along the internal surface. (Magnification: 6.5×.)

Figure 14.9   Zone of defects running along the internal surface.

CASE HISTORY 14.5

Industry:	Steel
Specimen Location:	Stack plate cooling section, blast furnace
Specimen Orientation:	Slanted
Environment:	Internal: Cooling water treated with scale inhibitor
	External: Refractory
Time in Service:	1 year
Sample Specifications:	Cast copper

Figure 14.10 shows the end profile of a sectioned stack plate with deep, irregularly shaped casting voids at the intersection of walls. Sectioning through these void zones revealed deep internal tunnel porosity (Fig. 14.11). When viewed under a low-power microscope, the contours of porous areas showed distinct solidification features (dendrites).

The voids are a casting defect resulting from insufficient feed of molten metal during the casting process. During freezing, shrinkage cavities are created due to the volume change that occurs when the metal passes from the liquid to the solid state. Additional liquid metal must be supplied to fill the cavities as the casting solidifies, or casting voids of the type illustrated may occur.

Figure 14.10 Cross section through a stack plate showing casting voids located at the junction of two walls (black spots at center of photograph).

Figure 14.11 Sectioning along the voids reveals the extent of the defect.

Weld Defects

The subject of weld defects is quite extensive, and complete coverage is well beyond the scope of this chapter. Therefore, this chapter will focus on specific types of weld defects of general concern in cooling water systems. Defects of seam-welded tubes are considered under material defects in Chap. 14.

Sound, fully satisfactory welds are the rule rather than the exception. As a joining technique, welding remains unrivaled as the method of choice for critical applications requiring superior strength and integrity. Carefully prepared welding procedures, faithfully followed by competent welders, consistently produce trouble-free welds. "Trouble-free" does not imply flawlessness, however. This is recognized by all major welding codes, which permit, within strict limitations, certain types of weld flaws.

A *weld* can be defined as the joining of two metals by fusing them at their interface. A metallurgical bond is formed that provides smooth, uninterrupted microstructural transition across the weldment. The weldment should be free of significant porosity and nonmetallic inclusions, form a smoothly flowing surface contour between the sections being joined, and be free of significant residual welding stresses.

A *weld defect* may be defined as a metallurgical or structural interruption in the weldment that significantly degrades the properties of the weld with respect to its intended use. Weld defects are not necessarily location specific. Any weld is a potential site of a defect. However, in cooling water systems, welded tube-to-tubesheet joints are often an area of concern.

Specific Weld Defects

Burnthrough

General description.   Burnthrough as discussed here specifically refers to the melting of tube metal in the vicinity of the weld such that a cavity is formed. If burnthrough is severe, a continuous channel may be produced that can cause leakage.

Locations.   Burnthrough in cooling water equipment may affect welded tube-to-tubesheet joints.

Critical factors.   Burnthrough may result from the use of excessive welding heat relative to the thickness of the tube.

Identification.   If accessible, defects from burnthrough may be visually identified as fused holes in the tube wall. Various nondestructive testing techniques, such as radiography and ultrasonics, may also detect this defect. The defect generally causes leakage soon after affected equipment is placed in service.

Elimination.   Careful use of appropriate welding procedures and techniques, as these relate to metal temperature, is necessary, especially when welding thin-walled tubes.

Galvanic corrosion

General description.   Galvanic corrosion refers to the preferential corrosion of the more reactive member of a two-metal pair when the metals are in electrical contact in the presence of a conductive fluid (see Chap. 16, "Galvanic Corrosion"). The corrosion potential difference, the magnitude of which depends on the metal-pair combination and the nature of the fluid, drives a corrosion reaction that simultaneously causes the less-noble pair member to corrode and the more-noble pair member to become even more noble. The galvanic series for various metals in sea water is shown in Chap. 16, Table 16.1. Galvanic potentials may vary with temperature, time, flow velocity, and composition of the fluid.

It is well known that galvanic corrosion may occur when two galvanically dissimilar metals are welded together or when the weld metal itself is sufficiently less noble than the metals it joins. However, the corrosion of the weld metal, as presented here, is a special case of general galvanic corrosion. It results from the manner in which some weld metals solidify. Weld metal may experience alloy segregation as the metal freezes. This microsegregation may result in marked differences in chemical composition within the weld, including the formation of a second phase, that can result in galvanic corrosion of the less-noble metal or phase. Susceptibility to corrosion is greatly increased if second phases form interconnected or continuous networks.

Locations. Galvanic corrosion of any type is most severe in immediate proximity to the junction of the coupled metals. Galvanic corrosion of weld metals is frequently microstructurally localized. The less-noble weld material will corrode away, leaving behind the skeletal remnants of the more-noble metal (Figs. 15.1 and 15.2).

Critical factors. Two factors are critical in galvanic corrosion of weld metal. The first is the existence of sufficient compositional differences

Figure 15.1 Preferential corrosion of the less noble of two phases in a weld bead (cross section). (Magnification: 50×.)

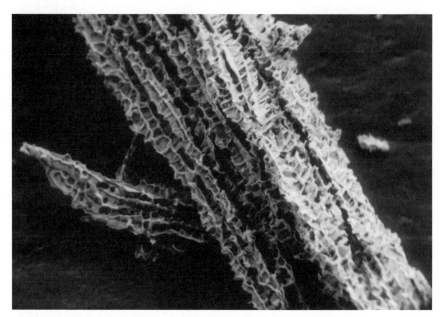

Figure 15.2 Skeletal remnants of noble phase. (Magnification: 200×, scanning electron microscope.)

through the weld metal. The second is exposure of the weld metal to a sufficiently aggressive environment.

Identification.   The primary identifying feature is confinement of metal loss to the weld bead (Fig. 15.3), although in advanced stages base metal immediately adjacent to the weld bead may also be affected. Note that this feature seems to distinguish galvanic corrosion of welds from other weld-related corrosion, such as weld decay, which preferentially attacks the immediately adjacent base metal (Fig. 15.4).

Occasionally, corrosion of this type produces large cavities covered by a thin outer skin of weld metal (Fig. 15.5). Even close examinations of such sites under a low-power microscope may fail to reveal the cavities. Compare Figs. 15.6 and 15.7. Generally, such sites are detected either by fluid leakage or by nondestructive testing techniques such as radiography and ultrasonics.

Elimination.   Recall that the critical factors governing galvanic corrosion of welds are the presence of substantial compositional differences within the weld metal and the exposure of such a weld to a sufficiently aggressive environment. If the aggressiveness of the environment cannot be sufficiently reduced, significant compositional differences within the weld metal must be avoided. This requires following proper

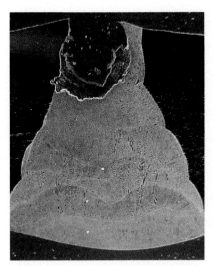

Figure 15.3 Corrosion confined to weld bead (cross section). (Magnification: 7.5×.)

Figure 15.4 Preferential attack of base metal adjacent to weld bead (top right—cross section). (Magnification: 50×.)

Figure 15.5 Cavity in weld bead covered by thin skin of weld metal (cross section). (Magnification: 7.5×.)

Figure 15.6   Weld bead site at which radiography indicated a large pit. (Magnification: 7.5×.)

welding techniques, protecting the molten weld metal from interaction with the environment, and choosing the correct weld-metal alloy.

Cautions.   Certain types of stainless steel welds are metallurgically designed to form two compositionally distinct phases to reduce the

Figure 15.7   Same site as Fig. 15.6 after probing with a pin. (Magnification: 7.5×.)

effects of weld-metal solidification shrinkage. Such welds must be used prudently when exposure to aggressive environments is anticipated. Also, high weld heat input has been linked to microstructural galvanic corrosion of welds (Case History 15.1).

Incomplete fusion

General description.   In incomplete fusion, complete melting and fusion between the base metal and the weld metal or between individual weld beads does not occur (Fig. 15.8). Incomplete fusion that produces crevices or notches at surfaces can combine with environmental factors to induce corrosion fatigue (Chap. 10), stress-corrosion cracking (Chap. 9), or crevice corrosion (Chap. 2). See Fig. 15.9.

Location.   Incomplete fusion may occur in any welded joint.

Critical factors.   The basic cause of incomplete fusion is failure to elevate the temperature of the base metal, or of the previously deposited weld metal, to the melting point. In addition, failure to flux metal oxides or other foreign substances adhering to metal surfaces properly may interfere with proper fusion.

Identification.   Incomplete fusion generally results in discontinuities along the side walls of a joint. When these discontinuities emerge at surfaces, they can be observed visually if accessible (Figs. 15.10 and 15.11). Defects of this type may also be detected by ultrasonics, radiography, magnetic particle inspection, and eddy-current testing.

Elimination.   The welding technique must be such that the melting points of the metals to be joined are reached and maintained until a

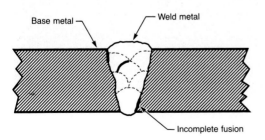

Figure 15.8 Common sites of incomplete fusion.

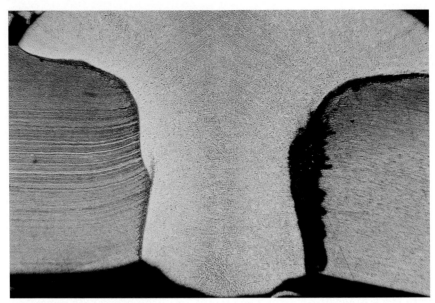

Figure 15.9 Cross section of stainless steel weld showing crevice corrosion along a site of incomplete fusion. (Magnification: 15×.)

Figure 15.10 Unfused seam along edge (top) of weld bead. (Magnification: 7.5×.)

Figure 15.11   Cross-sectional view of Fig. 15.10. (Magnification: 7.5×.)

metallurgical bond can be formed. This requires proper travel speed, proper electrode size, sufficient current, and proper electrode manipulation. It is also sound practice to clean the metal surfaces to be welded (note Case History 15.2).

Related problems.   See incomplete penetration below.

Incomplete penetration

General description.   Incomplete penetration describes the condition in which the weld fails to reach the bottom of the weld joint, resulting in a notch located at the root of the weld (Fig. 15.12). This critical defect can substantially reduce the intrinsic mechanical strength of the joint and can combine with environmental factors to produce corrosion fatigue (Chap. 10), stress-corrosion cracking (Chap. 9), or crevice corrosion (Chap. 2).

Location.   Incomplete penetration can occur at any welded joint.

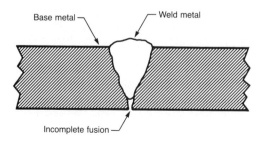

Figure 15.12 Incomplete penetration.

Critical factors.   Incomplete penetration is generally a consequence of insufficient heat to attain fusion temperatures along the weld surfaces at the bottom of the weldment. This is commonly caused by an unsatisfactory groove design for the welding process used. Other conditions leading to incomplete penetration are use of an electrode that is too large, inadequate welding current, or excessively high travel speed.

Identification.   If the notch left by incomplete penetration emerges at a visually accessible surface, visual examination, perhaps aided by magnetic-particle or liquid-penetrant techniques, may reveal the defect. Otherwise, ultrasonics, radiography, or eddy-current techniques may have to be used.

Elimination.   Proper joint design relative to the welding process used is important. Also, properly sized electrodes, proper travel speed, and adequate welding current are necessary.

Related problems.   See incomplete fusion above.

Laminations

General description.   Laminations are thin surfaces of nonmetallic substances, often oxides, sulfides, or silicates, that can run parallel to the surface of wrought metals such as plate, pipe, and tubes. They generally originate in the ingot and are expanded into thin sheets during subsequent hot-working processes. Laminations can cause defects or cracks in welded components (Fig. 15.13). Welding imposes stresses that can cause the lamination to spread apart. The crack formed in this manner may then propagate into the weld itself.

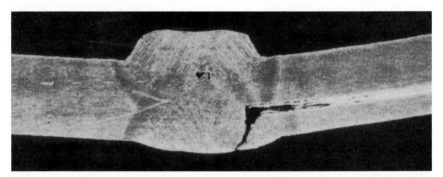

Figure 15.13 Cracking in weld deposit caused by lamination in steel base metal. (*Reprinted with permission from Helmut Thielsch,* Defects and Failures in Pressure Vessels and Piping, *New York, Van Nostrand Reinhold, 1965.*)

Locations.   Locations of defects of this type are essentially unpredictable but can occur wherever laminated metal is welded.

Critical factors.   The sole critical factor is the presence of a lamination of sufficient size to prevent the forming of a satisfactory weld.

Identification.   Defects in welds due to laminations may be difficult to identify based solely on a visual examination. Typically, cracks from welding laminated metal cause jagged, stepwise tears in the base metal. The crack may propagate into the weld itself.

Elimination.   Since laminations result from steel-making and steel-forming processes, little can be done to eliminate defects once they have survived quality inspections at the mill. If laminations are suspected, ultrasonics or radiography may disclose them. They may also be observed visually at cut ends of plate, pipe, or tubes if the cut intersects the lamination.

Porosity

General description.   Porosity refers to cavities formed within the weld metal during the solidification process. Such cavities may form due to decreased solubility of a gas as the molten weld metal cools or due to gas-producing chemical reactions within the weld metal itself. At times, cavities can form a continuous channel through the weld metal (worm holes, piping), resulting in leaks (Case History 15.3).

Locations.   Worm holes or piping may affect welded tube-to-tubesheet joints.

Critical factors.   In general, porosity is caused by the entrapment of gas during the welding process or during solidification of the weld metal. Surface contamination may provide a gas source during the welding operation.

Identification.   If accessible, welds having cavities may be visually observed by noting the pore formed where the cavity emerges at the surface (Fig. 15.14). Nondestructive testing techniques include radiography, eddy-current testing, ultrasonics, and possibly magnetic-particle or liquid-penetrant inspections. Note that the last two techniques are useful only if pores emerge at the surface. Leakage from defects of this type may occur soon after start-up of the equipment.

Figure 15.14   Pore formed by emergence of weld cavity at the surface. (Magnification: 15×.)

Elimination.   Porosity in general can be minimized by using clean, dry materials and by properly controlling weld current and arc length.

Slag inclusions

General description.   Slag inclusions are various nonmetallic substances that become entrapped in the weld during the welding process. Typically, the inclusions are located near the surface and along the sides of the weld (Fig. 15.15). The inclusions may form from reactions occurring in the weld metal or may be metal oxides present on the metal prior to welding. They may be isolated particles or may form relatively continuous bands.

Locations.   Slag inclusions are principally a result of improper welding techniques and therefore may be found wherever welds are present.

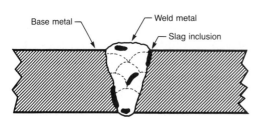

Figure 15.15   Common slag inclusion sites in a weld.

Critical factors.   Slag entrapment can occur if weld-metal temperature is too low or if solidification is too rapid.

Identification.   Slag inclusions will not be visually identifiable unless slag particles emerge at the weldment surfaces. Radiography, eddy-current testing, and ultrasonics are nondestructive testing techniques that can disclose slag inclusions.

Elimination.   Since slag is less dense than the weld metal, it will float to the surface if unhindered by rapid solidification. Therefore, preheating the components to be welded or high weld heat input may prevent slag entrapment.

Weld decay

General description.   The most common and best-known form of metallurgical weld defect in stainless steel is termed *weld decay.* Weld decay is disintegration in narrow zones of metal located adjacent and parallel to the weld, where sensitization occurs. *Sensitization,* in a welding context, refers to the formation of chromium carbides in the grain boundaries of the metals being joined as a consequence of the high temperatures produced during the welding process. Sensitization occurs in a specific temperature range that may be encountered along a pair of narrow zones parallel to the weldment. The formation of chromium carbides causes a severe depletion of dissolved chromium in the metal in an envelope surrounding each affected metal grain (Fig. 15.16). Since the corrosion resistance of stainless steel is directly linked to the concentration of chromium dissolved in the metal, the loss of chromium around each grain within these narrow zones renders them susceptible to various forms of degradation, such as corrosion and cracking, in a sufficiently aggressive environment.

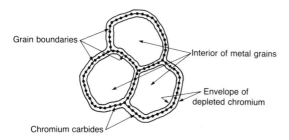

Figure 15.16 Schematic representation of sensitized stainless steel.

Stainless steel is susceptible to sensitization when it is heated to the range of 900 to 1550°F (480 to 850°C). Since any welding operation involving stainless steel will produce such temperatures in the metals being joined, it would appear that all welded stainless steel would sensitize. However, sensitization is a function of both time and temperature, occurring most rapidly at temperatures near 1250°F (675°C). Metals that cool rapidly through this temperature range will not sensitize. Consequently, thin metal sections, which cool rapidly, are less susceptible to sensitization than thick sections.

Note that sensitization will not result in weld decay in all environments. Stainless steels may be used in environments that do not require the full corrosion resistance of the alloy. In these cases, weld decay may not occur even though sensitization has taken place.

Locations.   Weld decay may affect welded stainless steels that have normal carbon contents and are not specifically inhibited for sensitization. Weld decay affects only the immediate weld area.

Critical factors.   The critical factors governing weld decay include the use of a sensitized stainless steel and the exposure of this metal to an environment that is sufficiently aggressive to cause degradation in the sensitized region.

Identification.   Weld decay characteristically produces distinct, narrow zones of disintegration immediately adjacent and parallel to the weldment (Fig. 15.4). Attack will occur on either side of the weld. The weld itself and surrounding metal will be unattacked. The corroded area frequently has a granular or sugary appearance and feel. This is due to corrosion along the boundaries of individual grains that causes them to be released from the metal surface. Affected metal may have lost its metallic ring.

In addition to the form of attack described above, sensitized welds are prone to pitting, stress-corrosion cracking in certain environments (see Chap. 9), and crevice corrosion (see Chap. 2).

Elimination.   The sensitization that leads to weld decay is a reversible process, and elimination can be achieved by appropriate heat treatment of the affected metal. Weld decay may also be avoided by specifying proper materials and welding techniques.

Apply heat treatment.   Applying heat treatment is a remedial rather than a preventive measure. Sensitized metal may be reheated to temperatures in the range of 1950 to 2050°F (1065 to 1120°C) for sufficient time to allow the chromium carbides to dissolve completely. If the metal is then rapidly cooled (quenched) through the sensitization range [900 to 1550°F (480 to 850°C)], chromium carbides will not have enough time to reform. Hence the sensitized structure is thermally erased and weld decay will not occur. Note the cautions below; this method has definite limitations. Consequently, successful elimination of the problem typically involves preventive techniques rather than remedial measures.

Specify low carbon grades of stainless steel. Since sensitization results from the formation of chromium carbides, one approach is to sufficiently reduce the level of carbon in the alloy. Reduction of the carbon level to 0.03% or less has been shown to be effective in preventing sensitization. The low carbon grade of 304 is designated 304L; 316 is 316L. Note the cautions below.

Specify stabilized grades of stainless steel. An alternative method to prevent chromium carbide formation is to charge the alloy with substances whose affinity for carbon is greater than that of chromium. These substances will react preferentially with the carbon, preventing chromium carbide formation and thereby leaving the chromium uniformly distributed in the metal. The carbon content of the alloy does not have to be reduced if sufficient quantities of these stabilizing elements are present. Titanium is used to produce one stabilized alloy (321) and niobium is used to provide another (347). Note the cautions below.

Alter welding methods. Sensitization is governed by both time and temperature. Consequently, factors such as the thickness of the components to be welded, the type of welding procedure, and the time required to perform the welding will influence the likelihood of inducing potentially damaging sensitization. Experience has indicated that electric arc welding, which characteristically produces intense heat over a short time, is the preferred technique for welding stainless steel that is not a low carbon or stabilized grade.

Cautions. Sensitization is a metallurgical condition. It can be identified by certain specialized nondestructive testing techniques or by destructive metallurgical examinations, but it cannot be identified by simple visual examination. It becomes visually apparent only after exposure of the sensitized metal to a sufficiently aggressive environment produces corrosion; that is, weld decay.

Note that low carbon or stabilized grades of stainless steel do not possess intrinsically greater corrosion resistance than their unadjusted counterparts. Their sole value in typical cooling water systems results from their resistance to sensitization and potential weld decay that can result when the metals are welded. It is therefore not economically justifiable to specify low carbon or stabilized grades of stainless steel for typical cooling water system components that are not to be welded.

Reheating and quenching sensitized stainless steel may not be practical in many cases. Note also that the quenching operation can induce substantial residual stresses and warpage.

Weld-root cracking

General description.   In weld-root cracking, cracks originate at the root of the weld (Fig. 15.17). Such cracks may propagate into the weld, through the weld, into adjacent components, or through a relatively brittle heat-affected-zone base metal.

Locations.   Weld-root cracking in cooling water equipment may affect welded tube-to-tubesheet joints but could potentially affect any welded joint.

Identification.   Weld-root cracks originate at the root of the weld and run longitudinally along the weld, perpendicularly to the base-metal surface and parallel to the axis of the weld. In general, they may be identified visually or by various nondestructive testing techniques such as radiography or ultrasonics. Failures from weld-root cracking may occur soon after start-up or after extended periods of successful service.

Elimination.   In multipass welding, care must be exercised to remove cracks in the initial pass, before subsequent passes are applied. Pre-

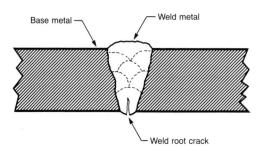

Base metal   Weld metal

Figure 15.17   Weld-root crack.

Weld root crack

cautions must also be taken to avoid creating stress-concentrating notches in welds and to keep residual welding stresses to a minimum.

Cautions.   Precise determination of the cause of weld-root cracks can be difficult. Determining correct repair procedures and preventive measures may require metallographic examinations and analyses.

Welding Stresses

General description

Stresses from welding result principally from the effects of differential thermal expansion and contraction arising from the large temperature difference between the weld bead and the relatively cold adjacent base metal. Shrinkage of the weld metal during solidification can also induce high residual stresses. Unless these residual stresses are removed, they remain an intrinsic condition of the weldment apart from any applied stresses imposed as a result of equipment operation.

Stresses resulting from welding are termed *residual* (or internal) *stresses*. Stresses resulting from equipment operation are called *applied* (or external) *stresses*. Residual stresses can be as high as the yield strength of the metal. Consequently, residual welding stresses can cause weld or adjacent base-metal cracking, even in the absence of applied stresses. However, the principal concern here is the interaction of high-weld stresses with the environment to produce other forms of cracking, such as stress-corrosion cracking. In addition, high-residual-welding stresses can act in combination with notches or other discontinuities in the weld to produce brittle fracture, especially if cyclically applied stresses are operating.

Although not commonly listed as a weld defect, high-welding stress nevertheless provides an essential ingredient to stress-corrosion cracking and other failures. It differs in an important respect from other types of weld defects in that stresses cannot be visually identified or revealed by conventional nondestructive testing techniques.

Locations

Stress-corrosion cracking can result from high-welding stresses in or immediately adjacent to the weld (Figs. 15.18 and 15.19).

The cracks illustrated in Fig. 15.18A resulted from weld metal-shrinkage along the long axis of the weld. The cracks in Fig. 15.18B were caused by shrinkage along the short axis of the weld. The cracking illustrated in Fig. 15.18D may be caused by lack of penetration.

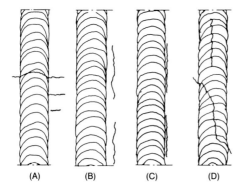

Figure 15.18 Examples of crack patterns due to stress-corrosion cracking and corrosion fatigue in butt welds. (*Reprinted with permission from Helmut Thielsch, Defects and Failures in Pressure Vessels and Piping, New York, Van Nostrand Reinhold, 1965.*)

(A)   (B)   (C)   (D)

Figure 15.19 shows various crack orientations that can occur in connection and attachment welds. Applied stresses from external loading of these components can add to the residual weld stresses, producing still higher stress loads. This can increase the susceptibility to stress-corrosion cracking and can affect orientation and location of crack paths.

Critical factors

Welding is a primary source of residual stresses principally because of large temperature differences between the weld bead and the rela-

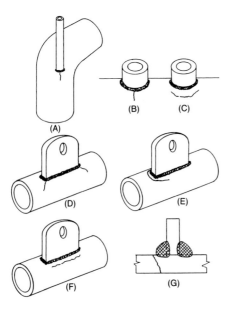

(B)   (C)

(A)

Figure 15.19 Examples of crack patterns due to stress-corrosion cracking and corrosion fatigue in branch connections and vessel or pipe support attachment welds. (*Reprinted with permission from Helmut Thielsch, Defects and Failures in Pressure Vessels and Piping, New York, Van Nostrand Reinhold, 1965.*)

(D)   (E)

(F)   (G)

tively cold adjacent base metal and because of the effects of weld-metal shrinkage during solidification.

Identification

Welding stresses are not visually observable. The common nondestructive technique available for determining residual stresses is x-ray diffraction.

Gross cracks may be visually observable. Nondestructive testing for the presence of cracks includes using dye penetrant, ultrasonics, and radiography. Determination of the cracking mechanism will require metallographic analysis.

Elimination

Residual stresses caused by large temperature differences between the weld bead and adjacent metal can be reduced by preheating the metals to be welded, especially if the sections are thick, or by stress-relief-annealing following the welding procedure. Proper welding techniques, especially the use of appropriate weld filler metals to minimize weld-metal shrinkage, can minimize residual stresses.

Cautions

When stress-relief-annealing 300 series stainless steel components, care must be taken to avoid slow cooling through the sensitization range (see "Weld Decay" in this chapter).

General Cautions

The importance of a competent, experienced weld inspector in evaluating weld defects cannot be overstated. Competent weld inspection requires, in part, distinguishing visually unappealing but serviceable welds from visually satisfying but defective welds. Such judgment requires knowledge of and experience in the techniques of welding, weld repair, and nondestructive testing, as well as service requirements, metallurgy, and defect interpretation. It is not uncommon to find that weld repairs made primarily for cosmetic reasons introduce features (such as residual stresses) that eventually result in weld failure. The mere presence of pores, slag inclusions, sensitization, and so on, in a weldment does not necessarily constitute sufficient grounds for rejecting the weld. A perfect weld is not a weld without flaws but a weld that will satisfactorily perform its intended function.

CASE HISTORY 15.1

Industry:	Utility
Specimen Location:	Distribution headers, recirculating cooling system
Specimen Orientation:	Horizontal, vertical
Environment:	Internal: Cooling water treated with corrosion inhibitors and sodium hypochlorite biocide, 75°F (24°C), 50 psi (345 kPa), pH 7–8
	External: Ambient, atmospheric
Time in Service:	2 years
Sample Specifications:	14 in. (35½ cm) diameter 316L stainless steel pipe; 308L weld filler metal

Operation of this cooling water system was intermittent, resulting in long periods (30 days) of no-flow conditions. After 2½ years, leaks were found at welded pipe junctions. Radiographic examinations revealed numerous additional deep corrosion sites at welds that had not yet begun to leak.

Several of the welded junctions were removed from the system for metallographic examination (Fig. 15.20). As can be seen from Fig. 15.20, the internal surface was covered with reddish and tan deposits and corrosion products. The metal surface itself retained a bright, metallic luster.

Close examination of the weld under a low-power stereoscopic microscope revealed small openings (Fig. 15.6). Probing these sites with a pin revealed a large pit that had been covered by a thin skin of weld metal. These sites contained fibrous, metallic remnants (Fig. 15.7). Examination under a scanning electron microscope further revealed the fibrous character of the material (Fig. 15.2) and also the convoluted shapes of the individual fibers (Fig. 15.21). Energy-dispersive spectrographic analysis of this material revealed the compositions in Table 15.1.

Figure 15.5 shows a transverse cross section through a pitted weld. Note the cavernous pit below a small opening in the surface skin.

The only significant corrosion observed in the entire system was confined to the weld beads along the internal surface. Corrosion occurred due to a microstructural galvanic couple formed between two distinct phases in the weld-bead microstructure (Figs. 15.22 and 15.23). The less-noble phase corroded away, leaving behind the skeletal remnants of the more-noble phase.

The triggering mechanism for the corrosion process was localized depassivation of the weld-metal surface. Depassivation (loss of the thin film of chromium oxides that protect stainless steels) can be caused by deposits or by microbial masses that cover the surface (see Chap. 4, "Underdeposit Corrosion" and Chap. 6, "Biologically Influenced Corrosion"). Once depassivation occurred, the critical features in this case were the continuity, size, and orientation of the noble phase. The massive, uninterrupted network of the second phase (Figs. 15.2 and 15.21), coupled with its exposed orientation along the surface of the weld bead, made this particular region of the weld bead susceptible to self-sustained galvanic corrosion once depassivation occurred. Research has demonstrated that the microstructural features of a large, continuous, surface-lying second phase can be produced by excessively high heat input during the welding process. Corrosion of this type could be avoided by reducing heat input during welding.

It is worth noting that corrosion failures essentially identical to the one presented here have been frequently attributed in the literature to microbiologically influenced corrosion. Whereas microbial masses may participate in the triggering mechanism by inducing localized depassivation, depassivation can occur under numerous environmental circumstances that have nothing to do with microbiological activity. In the laboratory, corrosion of this type has been induced simply by scratching (thereby depassivating) the surface of a weld having a susceptible microstructure. What is perhaps more crucial in mitigating this problem is that the diagnosis of microbiologically influenced corrosion ignores the critical importance of the second-phase microstructure as an essential galvanic driving force for the corrosion process. Once corrosion at such a site begins, it can continue in a self-sustained manner (autocatalytically), without microbial intervention, by spontaneous concentration of chloride ion and hydrogen ion by hydrolysis, a mechanism reported in the corrosion literature (see Chap. 2, "Crevice Corrosion").

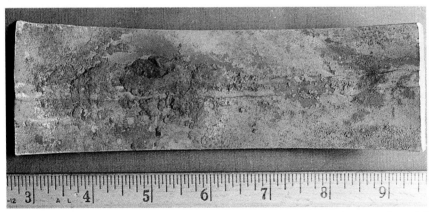

Figure 15.20   Internal surface of 14-in. diameter stainless steel pipe. The circumferential weld is in the center of the specimen.

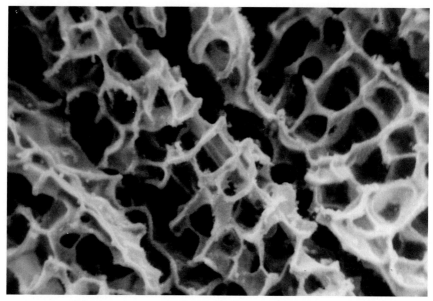

Figure 15.21 Convoluted contours of individual fibers. (Magnification: 1000×, scanning electron microscope.)

TABLE 15.1 **Chemical Compositions of Fibers and Weld Metal**

Element	Fibers (%)	Weld metal (%)
Iron	57	66
Chromium	33	22
Silicon	3	0.5
Nickel	2.5	10
Molybdenum	2	1.5

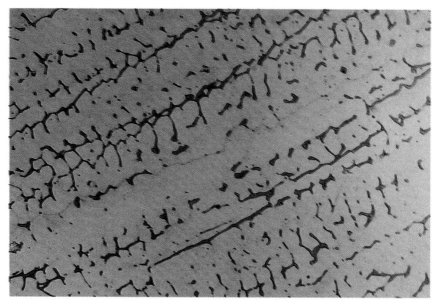

Figure 15.22   Phase 1: (noble) dark script. Phase II: light background. (Magnification: 500×, etched nitric acid.)

Figure 15.23   Preferential corrosion of Phase II; Phase I remains unaffected. (Magnification: 500×, unetched.)

CASE HISTORY 15.2

Industry:	Primary metals
Specimen Location:	Aluminum remelt furnace door frame
Specimen Orientation:	Slanted—immediately below arch
Environment:	Internal: Low-flow, treated (polymer, surfactant, biocide) cooling water, pH 8.5, 120–160°F (50–70°C), conductivity 800 μmhos
	External: Furnace heat
Time in Service:	4½ months
Sample Specifications:	Welded carbon steel

The section illustrated in Figs. 15.24 and 15.25 was removed from the door frame that supports the side-loading door to the furnace (Fig. 15.26). Failures of this type had been a chronic problem in the door frame area. The section had been fabricated by welding together three steel plates of ½ to ¾ in. (1.3 to 1.9 cm) thickness.

Exposure of the fracture surface revealed two distinct regions (Fig. 15.27). A relatively smooth region at the bottom of the J (top of photograph) is weld metal. The relatively rough surface that forms the trunk of the J is a fractured ¾-in. (1.9-cm) plate.

Close examination of these areas under a low-power microscope revealed smoothly rippled, spherical surfaces in the weld region and a chevron pattern that pointed back to the weld in the plate. Cross sections cut through the weld revealed substantial subsurface porosity and regions where oxidized surfaces prevented metallurgical bonding of the weld (Fig. 15.28).

It is clear from examination of the fracture surface and weld cross sections that the weld was improperly formed, resulting in an irregular plane of unbonded metal. The smoothly rippled, spherical contours in some regions of the fractured area are evidence of solidification of the weld metal along a free surface that was not in contact with the plate. Substantial porosity is apparent.

Microscopic examinations revealed iron oxide coatings on the plate that interfered with metallurgical bonding to the weld metal.

These weld defects not only substantially reduced the mechanical strength of the weld, the pores also formed stress-concentrating notches. Consequently, when the door frame was inadvertently struck during the scrap charging operation, a fracture initiated at the deficient weld and propagated rapidly through the ¾-in. (1.9-cm) plate, as revealed by the brittle appearance of the plate fracture and the directional chevron markings.

Figure 15.24   Section of cracked door frame.

Figure 15.25   Appearance of fracture along the external surface.

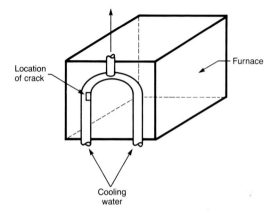

Figure 15.26 Schematic illustration of remelt furnace showing location of failure.

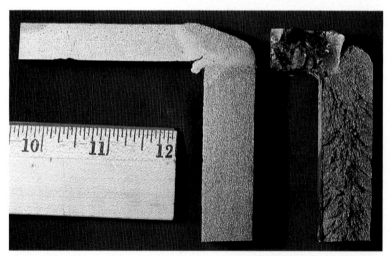

Figure 15.27 Fracture face (right) contrasted with corresponding, intact weldment. Note that the transition from a rough to a smooth fracture surface occurs abruptly at the junction of the weld metal with the base metal.

Figure 15.28 Cross section of fracture along the weldment showing subsurface porosity. Note that the fracture path runs through a large pore. (Magnification: 7×.)

CASE HISTORY 15.3

Industry:	Automotive
Specimen Location:	Oil cooler in an automobile engine
Specimen Orientation:	Vertical
Environment:	Internal: Lubricating oil, 300°F (150°C)
	External: Treated cooling water, 176°F (80°C)
Time in Service:	Unknown
Sample Specifications:	Welded, wrought aluminum

Pressure testing of the finned oil cooler in Fig. 15.29 revealed leaks. Examination of the interior of the cooler after sectioning in the vicinity of the leaks revealed a small cavity in the weld zone in the corner of some fins (Fig. 15.14) and porous areas inside the channel in the welded zone in other fins. Microstructural examinations of specimens cut through the sites revealed interconnected voids resulting from either shrinkage during solidification of the weld or lack of fusion of the base metal and weld metal.

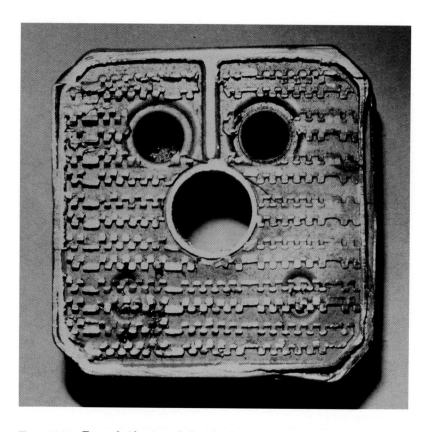

Figure 15.29   Top and side view of oil cooler. Fins at arrows.

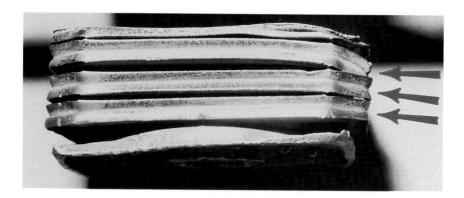

Galvanic Corrosion

General Description

Of the many types of corrosion that can affect cooling water systems, one of the most familiar is galvanic corrosion. The basic factors involved in galvanic corrosion are relatively simple and the concepts comparatively uncomplicated. Nevertheless, it is common to find galvanic corrosion improperly credited with the deterioration of one metal simply because a different metal is nearby. In addition, certain features of the galvanic corrosion process occasionally can become complex and subtle, making correct diagnosis in these cases a challenge.

The basic features involved in galvanic corrosion can be clearly illustrated by considering a simple dry cell (Fig. 16.1). A dry cell consists of a corrosion-resistant (noble) material and an active material (zinc in the conventional dry cell) embedded in a common, conductive environment (electrolyte). [Note that the noble material in a galvanic corrosion couple can be a conducting nonmetal, such as graphite in the case of a dry cell, or iron oxide (mill scale) or iron sulfide on steel in cooling water systems.] The noble and active materials are linked by an external electrical pathway, which can be opened or closed. When the circuit is closed (such as by switching on a flash light), the active metal begins to corrode and the noble material remains intact. The corrosion of the active metal furnishes electrons that flow to the noble material

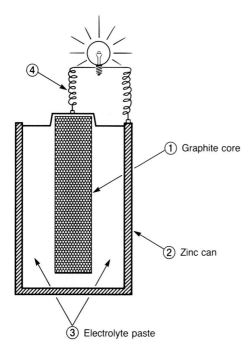

① Graphite core

② Zinc can

③ Electrolyte paste

Figure 16.1   Simple dry cell battery. Electrons are conducted along the external circuit (4), which physically connects the active (2) and noble (1) materials. An equivalent ionic countercurrent is conducted through the electrolyte (3), thereby completing the circuit.

through the external electrical pathway. A corresponding counterflow of positive ions moves through the electrolyte to complete the circuit. Hence, the basic features of a galvanic corrosion process are:

- An electrochemical interaction of two or more materials (materials 1 and 2 in Fig. 16.1) having a sufficiently distinct galvanic potential difference

- An electrolyte (3) common to both materials, through which an ionic current flows

- An electrically conductive pathway (4) physically linking the two materials

In any galvanic couple, the corrosion rate of the active material (anode) will typically increase, and the corrosion rate of the noble material (cathode) will typically decrease or cease altogether.

As corrosion proceeds, reaction by-products may form on the metal surfaces, creating a resistance to electrical exchanges at these surfaces. Consequently, the reaction rate diminishes correspondingly. If the corrosion reaction is stopped, a time-dependent recovery occurs; if the reaction is restarted, the initial corrosion rate is reestablished. This effect is often observed in conventional dry cells and automobile batteries.

The magnitude of this resistance phenomenon can vary from metal to metal. Hence, the rate and extent of galvanic corrosion for any particular couple may be less than anticipated when these effects are operating.

Galvanic corrosion typically involves two or more dissimilar metals. It should be recognized, however, that sufficient variation in environmental and physical parameters such as fluid chemistry, temperature (see Case History 16.3), flow velocity, and even variations in degrees of metal cold work can induce a flow of corrosion current even within the same metal.

Locations

Galvanic corrosion is location specific in the sense that it occurs at a bimetallic couple (Fig. 16.2). It is metal specific in the sense that, typically, corrosion affects the metal that has less resistance in the environment to which the couple is exposed. Hence, in principle, we would anticipate galvanic corrosion of relatively reactive metals wherever they are in physical contact with relatively noble metals in a sufficiently aggressive, common environment. Experience has shown, however, that all such couples do not necessarily result in unsatisfactory service. This is because of the interplay of various critical factors that influence galvanic corrosion. These critical factors are discussed in the next section.

The design of cooling water equipment such as cooling coils, heat exchangers, and condensers often involves dissimilar metal pairs in intimate contact, offering the possibility of galvanic corrosion. Tubes may be significantly noble with respect to the tube sheet metal or to baffles. Metallic inserts installed to prevent inlet-end erosion may induce galvanic corrosion of the less-noble tubes. Steel baffles may corrode where they contact copper tubes. Channels, heads, tube sheets, and gasket surfaces are targets of galvanic corrosion. Cast iron components such as floating head covers may galvanically corrode if coupled to relatively noble metals such as copper or copper-alloy tubes and tube

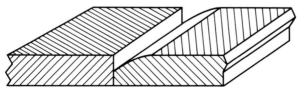

Figure 16.2   Galvanic corrosion along plane of bimetallic contact.

sheets. Interestingly, if the deterioration of the cast iron produces graphitic corrosion (see Chap. 17, "Graphitic Corrosion"), a potential reversal may occur in which the graphitically corroded cast iron becomes noble with respect to the copper or copper alloy, resulting in galvanic corrosion of these metals.

Galvanic corrosion may also occur by transport of relatively noble metals, either as particulate or as ions, to the surface of an active metal. For example, ions of copper, perhaps resulting from corrosion or erosion-corrosion at an upstream site, may be carried by cooling water to the surfaces of aluminum, steel, or even stainless steel components. If the ions are reduced and deposit on the component surfaces, localized galvanic corrosion may result.

Galvanic corrosion may occur at stainless steel welds if sensitization has taken place or if welding has produced unfavorable dissimilar phases (see Chap. 15, "Weld Defects," particularly Case History 15.1). These forms of microstructural galvanic corrosion do not involve the joining of two different metals in the usual sense.

Another form of microstructural galvanic corrosion, graphitic corrosion, is unique to gray and nodular cast irons. It may be encountered in cast iron pumps and other cast iron components. It is a homogeneous form of galvanic corrosion, not requiring connection to a different metal.

Homogeneous galvanic corrosion may also occur on the surface of steel components that are covered or partially covered with mill scale (magnetite, Fe_3O_4) or iron sulfide corrosion products. Both mill scale and iron sulfide are noble with respect to steel. Significant galvanic corrosion can occur where breaks or holidays in these corrosion products expose unprotected metal.

Sufficient temperature differences between sites on the same component can cause a galvanic current flow. In these cases, the site with the higher temperature is usually the corrosion site (see Case History 16.3). Galvanic corrosion of this form can potentially affect heat exchangers and refrigeration equipment.

Galvanic corrosion is more probable in all these various circumstances if the coolant is a conductive fresh water, brackish water, or sea water.

Critical Factors

The necessary conditions for galvanic corrosion are (1) a corrosive interaction of electrochemically dissimilar materials that are (2) exposed to a common conductive fluid and are (3) physically linked so that current can flow from one material to the other. Once these necessary conditions are met, both the magnitude of the corrosion and the determination of the anodic (corroded) and cathodic (protected) materials are affected by various environmental factors.

Galvanic corrosion in typical industrial cooling water systems is the net result of the interplay of these factors. Some factors may accelerate the corrosion process; others may retard it. In their approximate order of importance, the more influential factors are discussed below.

Galvanic potential

In simple terms, *galvanic potential* is related to the magnitude of the current induced by coupling dissimilar materials exposed to a common conductive fluid. The magnitude of the potential depends on the materials that are coupled and on the characteristics of the fluid to which the metals are exposed.

A galvanic series has been constructed that lists numerous industrial metals according to their galvanic potential in sea water. Table 16.1 is such a listing for metals often found in cooling water systems.

Strictly speaking, this arrangement is precise only for sea water under controlled laboratory conditions. In other conductive fluids under industrial conditions, the magnitude of the potential, and possibly the position of the metals on the list, could change. In typical cooling water environments, however, the order of the metals as listed for sea water would not be expected to change significantly.

Note that the table allows us to determine which metal in a couple will probably be noble and which will corrode. The table also suggests an approximate corrosion intensity. Hence, galvanic coupling of metals widely separated on the table, such as graphite and aluminum, would be expected to result in relatively rapid corrosion of the anodic member (aluminum in this example). Galvanic coupling of metals having little separation on the table, such as 304 stainless steel (passive) and cupronickel, would be expected to produce relatively low corrosion rates of the anodic member (cupronickel in this example). Bracketed metals would not be expected to produce significant galvanic corrosion if coupled with each other.

Table 16.2 summarizes the probable galvanic activity of various metals in sea water.

Conductivity of fluid

Exposure of the metals to a common fluid is necessary to complete the galvanic circuit. Galvanic corrosion does not occur on dry, coupled met-

TABLE 16.1   Galvanic Series of Selected Metals in Sea Water*

Noble

Graphite
Titanium

316 stainless steel (passive)
304 stainless steel (passive)

Monel (alloy 400)

70:30 cupronickel
90:10 cupronickel
400 series stainless steels (passive)
Silicon bronze
Manganese bronze
Copper
Admiralty brass
Red brass
Yellow brass
Naval brass
Muntz metal

316 stainless steel (active)
304 stainless steel (active)
400 series stainless steels (active)

Cast iron (not graphitically corroded)
Steel

Aluminum
Galvanized steel (note cautions)

Active

* Note that nonmetallic substances such as mill scale (Fe_3O_4) and iron sulfide (Fe_3S) are not listed. Nevertheless, on the surface of steel these substances can produce galvanic corrosion of the steel due to their relatively noble character.

als. Electrons flow through the metal to form one-half of the circuit (hence the requirement that the metals be electrically connected), and ions move through the fluid to provide the corresponding countercurrent. Consequently, the ionic conductivity of the fluid exercises a substantial influence on the galvanic corrosion process, affecting both its severity and extent. For example, galvanic corrosion in a fluid of relatively low conductivity may be moderate and confined to a narrow zone at the junction of the metals involved. In high-conductivity fluids, the corrosion will be more extensive (Fig. 16.2) and may be severe near the junction.

An interesting consequence of the influence of fluid conductivity is the observation that ionic currents can be diminished by geometry. For

TABLE 16.2   Sea Water Corrosion of Galvanic Couples

Metal considered		Galvanized steel	Aluminum 5052	Aluminum 3004	Aluminum 1100	Alclad	Aluminum 3003	Aluminum 6053	Aluminum 6061	Cadmium	Aluminum 2017	Aluminum 2117	Aluminum 2024	Low-carbon steel	Low-alloy steels	Cast iron	Low-alloy cast irons	12-14% Cr	16-18% Cr steel	Muntz metal	Manganese bronze	Naval brass	Yellow brass	Admiralty brass	Aluminum bronze	Red brass	Copper	Silicone bronze	70-30 Copper-nickel	Composition G bronze	Composition M bronze	Silver solder	70-30 nickel-copper	Cr-Ni stainless steel	Cr-Ni-Mo stainless steel	Graphite
Low-carbon steel	S	□	□	□	□	□	□	□	□	□	□	□	□	△	●	●	●	▲	▲	●	●	●	●	●	●	●	●	●	●	●	●	●	●	●	●	●
	E	○	□	□	□	□	□	□	□	□	□	□	□	△	●	●	●	▲	▲	▲	▲	▲	▲	▲	▲	▲	▲	▲	▲	▲	▲	▲	▲	▲	▲	▲
	L	○	○	○	○	○	○	○	○	○	○	○	○	△	■	■	■	■	■	■	■	■	■	■	■	■	■	■	■	■	■	■	■	■	■	■
Low-alloy steels	S	□	□	□	□	□	□	□	□	□	□	□	□	□	△	□	△	●	●	●	●	●	●	●	●	●	●	●	●	●	●	●	●	●	●	●
	E	○	□	□	□	□	□	□	□	□	□	□	□	○	△	○	△	●	●	▲	▲	▲	▲	▲	▲	▲	▲	▲	▲	▲	▲	▲	▲	▲	▲	▲
	L	○	○	○	○	○	○	○	○	○	○	○	○	○	△	○	△	■	■	■	■	■	■	■	■	■	■	■	■	■	■	■	■	■	■	■
Cast iron	S	□	□	□	□	□	□	□	□	□	□	□	□	△	●	△	●	●	●	●	●	●	●	●	●	●	●	●	●	●	●	●	●	●	●	●
	E	○	□	□	□	□	□	□	□	□	□	□	□	○	■	△	■	▲	▲	▲	▲	▲	▲	▲	▲	▲	▲	▲	▲	▲	▲	▲	▲	▲	▲	▲
	L	○	○	○	○	○	○	○	○	○	○	○	○	○	■	△	■	■	■	■	■	■	■	■	■	■	■	■	■	■	■	■	■	■	■	■
Low-alloy Cast iron	S	□	□	□	□	□	□	□	□	□	□	□	□	□	△	□	△	●	●	●	●	●	●	●	●	●	●	●	●	●	●	●	●	●	●	●
	E	○	□	□	□	□	□	□	□	□	□	□	□	○	△	○	△	▲	▲	▲	▲	▲	▲	▲	▲	▲	▲	▲	▲	▲	▲	▲	▲	▲	▲	▲
	L	○	○	○	○	○	○	○	○	○	○	○	○	○	△	○	△	■	■	■	■	■	■	■	■	■	■	■	■	■	■	■	■	■	■	■
12-14% Cr steel	S	□	□	□	□	□	□	□	□	□	□	□	□	□	□	□	□	△	●	●	●	●	●	●	●	●	●	●	●	●	●	●	●	●	●	●
	E	□	□	□	□	□	□	□	□	□	□	□	□	□	□	□	□	△	■	▲	▲	▲	▲	▲	▲	▲	▲	▲	▲	▲	▲	▲	▲	▲	▲	●
	L	○	○	○	○	○	○	○	○	○	○	○	○	○	○	○	○	△	■	■	■	■	■	■	■	■	■	■	■	■	■	■	■	■	■	■
16-18% Cr steel	S	□	□	□	□	□	□	□	□	□	□	□	□	□	□	□	□	□	△	●	●	●	●	●	●	●	●	●	●	●	●	●	●	●	●	●
	E	□	□	□	□	□	□	□	□	□	□	□	□	□	□	□	□	□	△	▲	▲	▲	▲	▲	▲	▲	▲	▲	▲	▲	▲	▲	▲	▲	▲	●
	L	○	○	○	○	○	○	○	○	○	○	○	○	○	○	○	○	○	△	■	■	■	■	■	■	■	■	■	■	■	■	■	■	■	■	■
Austenitic Cr-Ni stainless steel	S	□	□	□	□	□	□	□	□	□	□	□	□	□	□	□	□	□	□	●	●	●	●	●	●	●	●	●	●	●	●	●	●		△	●
	E	□	□	□	□	□	□	□	□	□	□	□	□	□	□	□	□	□	□	■	■	■	■	○	○	○	■	■	■	■	■	■	■		△	■
	L	○	○	○	○	○	○	○	○	○	○	○	○	○	○	○	○	○	○	△	△	△	△	○	○	○	△	△	△	△	△	△	△		△	●
Austenitic Cr-Ni-Mo stainless steel	S	□	□	□	□	□	□	□	□	□	□	□	□	□	□	□	□	□	□	●	●	●	●	●	●	●	●	●	●	●	●	●	●	□	△	●
	E	□	□	□	□	□	□	□	□	□	□	□	□	□	□	□	□	□	□	■	■	■	■	○	○	○	■	■	■	■	■	■	■	○	△	●
	L	○	○	○	○	○	○	○	○	○	○	○	○	○	○	○	○	○	○	△	△	△	△	○	○	○	△	△	△	△	△	△	△	△	△	▲

□ The corrosion of the metal under consideration will be reduced considerably in the vicinity of the contact.
○ The corrosion of the metal under consideration will be reduced slightly.
△ The galvanic effect will be slight with the direction uncertain.
■ The corrosion of the metal under consideration will be increased slightly.
▲ The corrosion of the metal under consideration will be increased moderately.
● The corrosion of the metal under consideration will be increased considerably.

S Exposed area of the metal under consideration is small compared with the area of the metal with which it is coupled.
E Exposed area of the metal under consideration is approximately equal to that of the metal with which it is coupled.
L Exposed area of the metal under consideration is large compared to that of the metal with which it is coupled.

SOURCE: *Metals Handbook,* vol. 13, 9th ed., 1987, p. 773.

instance, galvanic corrosion is reduced around a bend in a tube. This is apparently due to increased resistance to current flow.

Area effect

Most galvanic corrosion processes are sensitive to the relatively exposed areas of the noble (cathode) and active (anode) metals. The corrosion rate of the active metal is proportional to the area of exposed noble metal divided by the area of exposed active metal. A favorable area ratio (large anode, small cathode) can permit the coupling of dissimilar metals. An unfavorable area ratio (large cathode, small anode) of the same two metals in the same environment can be costly.

Correct application of this principle can lead to what would appear to be peculiar recommendations. For example, if just one member of a couple is to be coated, it should be the noble member. Most coating systems leave holidays or tiny openings where the metal is exposed. If the active metal is coated, the area of exposure at the holidays can be quite small compared to the area of the noble metal, resulting in an unfavorable area ratio. On the other hand, if the noble metal is coated, the holidays provide a small cathodic area and hence a highly favorable area ratio with respect to the active metal. Similarly, if dissimilar metal fasteners must be used, they should be noble relative to the metals being fastened (see Case History 16.1).

Surface effects

The galvanic series in Table 16.1 is generally useful for predicting the tendency toward galvanic corrosion between coupled metals. The arrangement of this series is, however, based on data generated under controlled laboratory conditions on clean, bare metals.

In industrial situations, metal surfaces may vary from clean to completely covered with deposits and corrosion products. These coverings may interfere with and retard the exchange of current on the fluid side of the circuit. Since the electron flow through the coupled metals must be equivalent to the ionic flow through the fluid, reduction in flow on the fluid side will result in a corresponding decrease in electron flow through the metals; in other words, the corrosion rate will diminish or possibly cease. Furthermore, it is not uncommon for films of retarding substances to form on the metal surfaces as a consequence of galvanic corrosion itself. Hence, initial corrosion rates may diminish to a fraction of their original value as the film forms and develops.

Fluid velocity

The possible effects of fluid velocity on galvanic corrosion are sometimes overlooked. Fluid velocity can affect the apparent potential of metals in a given environment. Depending on the environment, a metal under the influence of relatively rapid flow may assume either a more noble or a more active character than that indicated by the galvanic series. Occasionally, this shift in potential may result in galvanic corrosion that would not occur under stagnant or low-flow conditions.

Identification

Avoid the common predilection to diagnose as galvanic corrosion the deterioration of one metal simply because a second metal is nearby.

This error can generally be avoided if the basic elements of the galvanic couple, as outlined under "General Description," are duly recognized. Hence, corrosion of one member of a couple when both metals are galvanically similar is probably not due to galvanic corrosion (note "Cautions" with respect to stainless steels). Corrosion of one member of a couple exposed to a common, nonconducting fluid is not galvanic corrosion, since current must flow through the *fluid* to complete the galvanic circuit. Corrosion of one metal that is not coupled to a dissimilar metal is not galvanic corrosion, since a common current must pass through *both metals* to complete the galvanic circuit.

Galvanic corrosion will generally have the following features:

- The more active of the coupled metals will corrode.
- Corrosion of the noble metal will be slight or nonexistent, even though it would corrode in the given environment if it were not coupled to the active metal.
- In general, corrosion of the active metal will be most severe at its junction with the noble metal (Fig. 16.2) and will decrease with increased distance from the junction.
- The extent of the attack will vary depending on the conductivity of the fluid; the greater the conductivity, the more extensive the corrosion.

Elimination

The nature of galvanic corrosion is such that successful avoidance generally requires implementing preventive rather than corrective techniques. Therefore, consideration of galvanic corrosion problems must be integrated into the design of equipment. Corrective techniques applied to existing equipment can be expensive and less than satisfactory.

Consideration of the basic elements characteristic of the galvanic corrosion process, as discussed above, points to the principles of sound preventive techniques. Since a galvanic potential difference is the driving force for corrosion: reducing the magnitude of this difference can reduce or prevent galvanic corrosion.

Also, since the corrosion rate is directly influenced by the magnitude of the galvanic current flowing through the electrolyte and metals, increasing the resistance to current flow in the electrolyte and/or the metal will stifle galvanic corrosion. Similarly, galvanic corrosion can be prevented altogether by breaking the electrical circuit through the use of nonconductive materials at the junctions of the metals. Finally, the importance of area ratios should not be overlooked in equipment design, since favorable ratios may actually permit the use of materials having significantly dissimilar galvanic potentials.

Based on these considerations, specific preventive or corrective steps may be taken to minimize or eliminate galvanic corrosion.

Preventive techniques

1. When possible, avoid coupling materials having widely dissimilar galvanic potentials. If this cannot be avoided, make use of favorable area ratios by giving the active metal a large exposed area relative to the noble metal. For example, copper or copper-based alloy tubes may be joined to a steel tube sheet. Because of the favorable area ratio in this case, a relatively inexpensive steel tube sheet may be intentionally substituted for a bronze or a brass tube sheet if thickness specifications allow for a small amount of galvanic corrosion of the steel.

   Fasteners, such as bolts and rivets, that are galvanically dissimilar to the materials being fastened should be relatively noble to these materials.

2. Completely insulate the materials from one another at all junctions exposed to a common fluid. Insulation involves using nonconductive washers, inserts, sleeves, and coatings.

3. Specify low carbon or inhibited grades of stainless steel (304L, 316L, 321, or 347) for welded assemblies that will be exposed to aggressive environments. These grades prevent sensitization of the stainless steel base metal adjacent to the welds. Sensitized stainless steel is susceptible to homogeneous galvanic corrosion in sufficiently aggressive environments.

4. Specify appropriate welding techniques for stainless steels that may form more than one phase in the weldment (note Case History 15.1).

5. If materials widely separated on the galvanic series must be joined, avoid threaded connections. Brazing or welding is preferable.

6. If galvanically incompatible materials are to be used, design the active material component so that easy replacement is possible, or allow for anticipated corrosion by appropriately increasing its thickness.

Corrective techniques

1. Completely insulate the materials from one another at all junctions exposed to a common fluid. Insulation involves using nonconductive washers, inserts, sleeves, and coatings.

2. Alter the chemistry of the common fluid to render it less conductive and/or less corrosive. Generally, water corrosivity increases with an increase in temperature and oxygen content and a decrease in pH. Inhibitors may be effective. Note that in mixed-metal systems, higher dosages of inhibitors may be required than would be necessary in single-metal systems in the same environment.

3. Coat both materials or the noble material. *Do not coat just the active material* (see "Area effects" above). Coatings should be maintained, especially the one covering the active member of the couple. Use of cathodic protection systems in association with coatings is often recommended.

4. Use cathodic protection techniques.

5. Metal ions from corrosion of relatively noble equipment may plate out on more active metals downstream, setting up highly localized galvanic cells. This situation can be especially troublesome, for example, when copper ions are conveyed to aluminum, zinc-coated steel, carbon steel, and even stainless steel equipment. Such ions can be intercepted and scavenged by passing the fluid containing the ions over metal scraps or through easily replaceable pipes fabricated from metals that will quickly displace them. For example, aluminum scraps or replaceable tubing will readily scavenge copper ions.

Cautions

Avoid the common tendency to attribute to galvanic corrosion the deterioration of one metal simply because another metal is nearby. The conditions necessary for galvanic corrosion are specific, and all must be operating simultaneously for it to occur. These conditions are outlined in the "General Description" section of this chapter.

Carefully note the dual positions of stainless steels in the galvanic series (Table 16.1) corresponding to their active and passive states. Passive stainless steel (its normal condition in well-aerated, flowing water) is quite noble and can potentially induce galvanic corrosion on aluminum, carbon steel, and even copper and copper alloys. In its active condition, however, stainless steel is less noble than copper and may suffer galvanic corrosion if coupled to this metal or its alloys. (Stainless steel may lose its passivity—assume an active state—under low-flow or stagnant conditions and when oxygen and oxidizing ions are depleted, especially if chloride ions are present.)

Note also that a galvanic couple can be established between passive regions and active regions of the same stainless steel component. For example, active sites may form at crevices or beneath deposits, resulting in rapid corrosion rates, especially if the surrounding passive metal has a relatively large area (unfavorable area ratio). This caution applies to other passive metals, such as aluminum, as well.

Similarly, graphitically corroded cast iron (see Chap. 17) can assume a potential approximately equivalent to graphite, thus inducing galvanic corrosion of components of steel, uncorroded cast iron, and copper-based alloys. Hence, special precautions must be exercised when dealing with graphitically corroded pump impellers and pump casings (see "Cautions" in Chap. 17).

Remember that the galvanic series was constructed from laboratory data using sea water as the exposure fluid. When there is a question about galvanic corrosion tendencies in actual industrial environments involving fluids substantially different from sea water, appropriate testing of candidate metals in these fluids may be warranted.

Aluminum components are sensitive to ions of heavy metals, especially copper. To avoid localized galvanic corrosion of the aluminum by metallic copper reduced from copper ions, care must be exercised to prevent heavy metal ions from entering aluminum components. Note the recommendations under "Elimination."

The galvanic potential of metals can vary in response to environmental changes such as changes in fluid chemistry, fluid-flow rate, and fluid temperature. For example, at ambient temperatures steel is noble to zinc (as in galvanized steel). In waters of certain chemistries, however, a potential reversal may occur at temperatures above 140°F (60°C), and the zinc becomes noble to the steel.

An interesting effect is sometimes observed when cupronickels are galvanically coupled to less noble materials. The corrosion rate of the active metal is increased and the corrosion rate of the cupronickel is diminished, as expected. The diminished corrosion rate of the cupronickel can, however, diminish its fouling resistance since reduced production of copper ions lowers toxicity to copper-ion-sensitive organisms.

Care must be exercised when installing stainless steel inserts in the inlet or exit end of copper or copper-alloy tubes, since galvanic corrosion can occur at the tube-insert junction.

Caution is also necessary when changing the metallurgy of tubes in a heat exchanger. Choosing more noble tubes may induce galvanic corrosion of the tube sheet. Problems can result if the tube sheet is not thick enough to accommodate the increased corrosion rate.

CASE HISTORY 16.1

Industry:	Municipal cooling system
Specimen Location:	Threaded fastener
Specimen Orientation:	Horizontal
Environment:	Treated cooling water, alkaline program, pH 8.5–9.5, 50–105°F (10–40°C), total hardness 150 ppm, chlorides 70 ppm, calcium 90 ppm
Time in Service:	7 months
Sample Specifications:	3½ in. (8.9 cm) outer diameter, threaded coupling fabricated from cast zinc

Threaded fasteners coupled heaters to the basins of cooling towers. (The heaters were used to prevent freezing of cooling towers during the winter.) Failure of the fasteners was extensive, affecting every coupling in each of several towers.

The fasteners contacted sheets of stainless steel. Threaded regions of the fasteners that were exposed to water were severely wasted. Threaded regions that were not exposed remained intact. Some exposed fastener surfaces were coated with plastic. These areas were also unaffected.

Corrosion of the fasteners occurred due to their galvanic interaction with passive stainless steel. Deterioration was rapid because of the unfavorable area ratio formed by the large areas of stainless steel and the small area of the fasteners, which was further reduced by the incomplete plastic covering overexposed fastener surfaces.

Note that zinc anodes are often used to protect steel and other relatively noble metals cathodically. In this case, the fasteners were acting as unintentional sacrificial anodes, protecting the stainless steel. Simple solutions to the problem would be to insulate the fasteners from the stainless steel electrically or to use stainless steel fasteners.

CASE HISTORY 16.2

Industry:	Chemical process—plastics
Specimen Location:	Inner ring of cooling system following the extruder
Specimen Orientation:	Axis of ring is horizontal
Environment:	Treated cooling water 90–155°F (32–68°C), 35 psi (241 kPa)
Time in Service:	12 years
Sample Specifications:	15 in. (38.1 cm) diameter aluminum ring; the ring has a rectangular cross section measuring 1 × 2 in. (2½ × 5 cm); the ring is fitted with a series of brass inlet and outlet nozzles

Failures of the inner ring of the extruder cooling system were occurring with increasing frequency. Six failures had occurred over a 2-month period.

A half-section of a ring is shown in Fig. 16.3. The brass nipples have been removed, but two insertion holes are apparent at the center of the ring.

An irregular trough of metal loss is apparent along the circumference of the ring (Fig. 16.4). Metal loss is severe near the nozzle holes (Fig. 16.5). The corroded zone is covered with light and dark corrosion products and deposits. Analysis of these revealed substantial quantities of copper and zinc. Microscopic examinations revealed exfoliation of the aluminum ring in corroded regions.

The large amounts of copper and zinc found on corroded surfaces distant from the brass nozzles indicate corrosion of brass components in other regions of the cooling system. Ions of copper and zinc were transported via the cooling water to the aluminum ring, where metallic copper was deposited by a replacement reaction with the aluminum. The reduced metallic copper then established a galvanic couple with the aluminum ring, producing additional loss of aluminum. Deposition of the zinc was apparently a secondary cathodic reaction.

The more severe metal loss adjacent to the brass nozzles is apparently due to a direct galvanic interaction between the aluminum ring and the brass nozzles. The type of exfoliation observed microscopically in corroded areas is consistent with galvanic attack.

Complete and correct diagnosis of failures will account for all observed evidence. In this case, although some questions remain unanswered, it is judged that galvanic corrosion exercised the predominant influence.

Figure 16.3   Section of inner ring of extruder cooling system.

Figure 16.4   Appearance of corrosion along circumference of ring.

Figure 16.5   Severe deterioration adjacent to nozzle holes where brass nozzles had been inserted.

CASE HISTORY 16.3

Industry:	Pulp and paper
Specimen Location:	Turbine condenser tube
Specimen Orientation:	Horizontal
Environment:	Internal: Cooling water
	External: Steam and condensate
Time in Service:	Unknown
Sample Specifications:	⅞ in. (2.2 cm) outside diameter, 90:10 cupronickel

Figure 16.6 illustrates the internal surface of a section removed from the center of a tube in the second pass. It is apparent from the external surface (Fig. 16.7) that the large, irregular perforation is located along the edge of the contact plane of the tube with a baffle.

Close visual examinations revealed that the perforation originated on the internal surface along the bottom of the tube where it rested on the steel baffle. Apparently, tube metal temperatures were higher in the area of contact with the baffle than at adjacent locations. The hotter tube metal at the baffle established a thermogalvanic cell with the surrounding cooler metal, causing metal wastage in the hotter region.

Figure 16.6   Perforation in tube wall originating on the internal surface.

Figure 16.7 External surface at the same location as Fig. 16.7. Note location of baffle contact surface relative to the perforation.

Graphitic Corrosion

General Description

Graphitic corrosion has two distinct features that are useful in distinguishing it from other forms of corrosion. First, it affects an unusually limited number of metals; the only metals commonly affected are gray cast iron and nodular cast iron. Second, metal that has experienced graphitic corrosion may retain its original appearance and dimensions. Consequently, graphitic corrosion frequently escapes detection.

The basic mechanisms involved in graphitic corrosion are familiar and easily understood. Hence, remedial and preventive measures are relatively simple to implement. Although commonly categorized as a form of dealloying, graphitic corrosion has much in common with galvanic corrosion.

The basic mechanism causing graphitic corrosion is easily understood, but some information is required on the microstructure of cast iron. The microstructural feature of principal importance is the distribution of flakes (gray cast iron) or spheroids (nodular cast iron) of graphite embedded in a matrix of iron (Fig. 17.1). Since graphite is highly noble from a corrosion standpoint and since it is in intimate physical contact with the relatively innoble iron matrix, it is easy to conceive of the formation of a microstructural galvanic couple between the graphite and iron if both are exposed to the same sufficiently aggressive environment. Such a situation can occur at exposed surfaces of the casting where graphite flakes emerge. A microminiature battery develops that is mechanistically identical to the old carbon–zinc dry cell (Fig. 17.2). The noble graphite remains intact while the iron corrodes (Fig. 17.3), the corrosion being driven by the potential difference generated by the galvanic couple and the electrolyte environment.

It is interesting to note that the iron is not removed but is transformed in situ to iron oxide. What is even more remarkable is that the volume change anticipated when iron is converted to iron oxide does not occur. Consequently, surface contours of the corroded metal remain intact, even to the point of retaining fine detail such as machining marks.

Graphitic corrosion is a slow corrosion process, typically requiring many years to effect significant damage. Complete penetration of thick cross sections has, however, occurred in as little as 2 years in adverse environments. On the other hand, cast iron components can be found in use in Europe after 160 years of service. Although graphitic corrosion causes a substantial reduction in mechanical strength, it is well known that corroded cast iron, when sufficiently supported, may remain serviceable when internal pressure is low and shock loads are not applied.

Note two cautions: First, the term *graphitic corrosion* should not be used interchangeably with the term *graphitization*. Graphitization, a

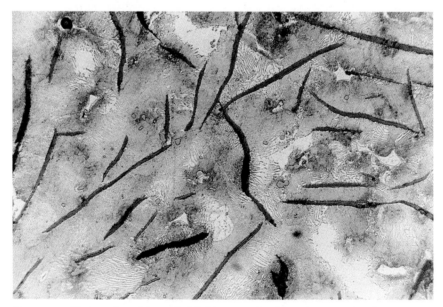

Figure 17.1   Flakes of graphite embedded in a matrix of iron (gray cast iron).

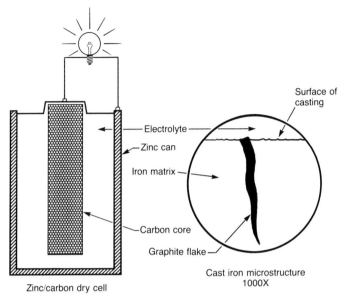

Surface of
casting

Electrolyte

Zinc can

Iron matrix

Carbon core

Graphite flake

Cast iron microstructure
1000X

Zinc/carbon dry cell

Figure 17.2 Mechanistic correlation between dry cell and graphitic corrosion.

Figure 17.3 Graphite flakes surrounded by graphitically corroded iron. Bright background is uncorroded metal. (Magnification: 1000×.)

distinctly different phenomenon, is the spontaneous conversion of iron carbide to elemental iron and elemental carbon (graphite) that can occur in carbon-containing irons and steels at temperatures above 850°F (455°C). Second, the literature frequently states that nodular cast iron is immune to graphitization. Although it is less prone than gray cast iron to serious graphitic corrosion, the basic microstructural features required for graphitic corrosion exist in nodular irons, and cases of graphitic corrosion of this metal have been observed by the authors.

Locations

The occurrence of graphitic corrosion is not location specific, other than that it may occur wherever gray or nodular cast iron is exposed to sufficiently aggressive aqueous environments. This includes, and is common to, subterranean cast iron pipe, especially in moist soil (Case History 17.1). Cast iron pump impellers and casings are also frequent targets of graphitic corrosion (Case Histories 17.2 through 17.5).

Critical Factors

Two critical factors govern graphitic corrosion:

1. A susceptible cast iron microstructure
2. Exposure of the metal to an environment that is sufficiently aggressive to generate a potential difference capable of driving a galvanic corrosion reaction between the graphite and the iron

Experience has demonstrated that graphitic corrosion is favored by relatively mild environments such as soft waters, waters having a slightly acidic pH, waters containing low levels (as little as 1 ppm) of hydrogen sulfide, and brackish and other high-conductivity waters. It should not be inferred from this that gray or nodular cast irons are immune to more aggressive environments but rather that graphitic corrosion is less likely in them. In more aggressive environments, corrosion may indeed occur but may be manifested as general metal loss rather than as graphitic corrosion. Moist soils, especially those containing sulfates, will frequently produce graphitic corrosion of unprotected gray and nodular cast iron. Stray currents have also been identified as causes of graphitic corrosion in subterranean pipelines. Finally, there is evidence that stresses may foster localized, relatively rapid graphitic corrosion.

Identification

As noted earlier, complete graphitic corrosion can occur without effecting dimensional changes in the component and without altering fine surface details such as machining marks and identification numbers. This can make mere visual identification difficult. Graphitically corroded metal may, however, change color from gray to black.

Affected metal is converted to a very soft material that can be dislodged with a sharp metal probe or cut with a knife. Graphitically corroded metal frequently has an oily, slippery feel, especially after surface deposits are scraped away. Because of its graphite content, chunks of corroded metal can be used as writing implements in the absence of a pencil (Fig. 17.4). Graphitically corroded metal will produce a dull thud rather than a metallic sound when struck with a hammer.

On subterranean pipeline, look for graphitic corrosion on the very bottom of the line where it rests on the backfill. When graphitic corrosion occurs under these conditions, the affected region may be a narrow zone running along the pipe bottom over some distance (Case History 17.1).

Failure of graphitically corroded cast iron will yield a brittle, thick-walled fracture. Fracture faces through the graphitically corroded region will be black and nonmetallic (Fig. 17.5). Cross sections cut through graphitically corroded regions will readily show bright, intact metal surrounded by a soft, black, corroded area (Fig. 17.6).

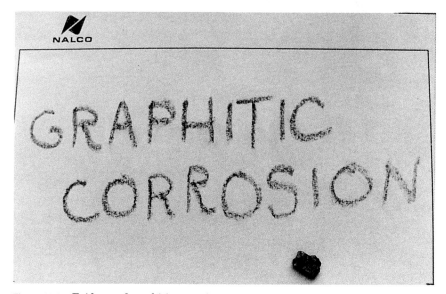

Figure 17.4 Evidence of graphitic corrosion.

Figure 17.5   Brittle fracture through pipe. The gray material is graphitically corroded cast iron. Unaffected pipe wall metal is orange from normal rust.

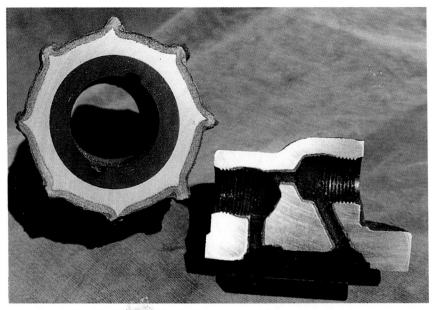

Figure 17.6   Black, graphitically corroded metal at the surface surrounding bright, unaffected metal.

Elimination

Control of graphitic corrosion can be effected by gaining control of the critical factors that govern it.

Susceptible microstructure

Altering material or microstructure is a preventive rather than a remedial technique and therefore cannot be applied to existing equipment. When graphitic corrosion is anticipated, consideration should be given to specifying alternate materials. Nodular cast iron is less prone to serious graphitic corrosion than other cast irons, but it is not immune. White cast iron, which is essentially free of graphite, is immune to graphitic corrosion. Corrosion-resistant cast irons containing chromium, nickel, or silicon are essentially immune to graphitic corrosion. Austenitic cast irons are also immune, as are cast steels.

Exposure to an adverse environment

Two approaches exist to gain control of this factor: altering the environment and isolating the metal from the environment. These are remedial measures and can be implemented during or after installation of the equipment.

- *Alter the environment.*   The key here is to make alterations to the environment that will render it nonaggressive, not only from the standpoint of graphitic corrosion but also with respect to other corrosion mechanisms. For example, since graphitic corrosion can be caused by exposure to slightly acidic waters, a pH adjustment (neutralization) may provide sufficient protection. Since sulfates contribute to graphitic corrosion, coal ashes should be avoided as backfill for subterranean pipelines. Also, backfill should be compacted to exclude air (oxygen) that may contribute to graphitic corrosion. These are simple examples of environmental alterations. Many other adjustments may be possible once pertinent environmental factors are identified (see "Critical Factors" above).

- *Isolate the metal from the environment.*   If environmental alteration is impossible or uneconomical, it may be appropriate to separate the metal from the environment. This is commonly the economically justifiable approach. It generally involves applying an appropriate protective coating to the metal. External coatings must be resistant to cold flow, especially in subterranean applications where the weight of a pipe resting on hard backfill may cause the coating to flow, opening holidays where graphitic corrosion can initiate. It is common practice to use cathodic protection in conjunction with coatings. Internal surfaces can be lined with cement mortar.

Cautions

It is important to note that graphitically corroded metal may be overlooked in a simple visual inspection. Appropriate use of a sharp probe or hammer should be helpful in the identification.

Note also that graphitic corrosion may occur preferentially in poorly accessible areas, such as the bottom of pipelines. Trouble-free service of cast iron components does not necessarily indicate that all is well, since components suffering severe graphitic corrosion may continue to operate until an inadvertent or intentional (e.g., pressure-testing) shock load is applied. At this point massive, catastrophic failures can occur.

Graphitically corroded cast irons may induce galvanic corrosion of metals to which they are coupled due to the nobility of the iron oxide and graphite surface. For example, cast iron or cast steel replacement pump impellers may corrode rapidly due to the galvanic couple established with the graphitically corroded cast iron pump casing. In this or similar situations, the entire affected component should be replaced. If just one part is replaced, it should be with a material that will resist galvanic corrosion, such as austenitic cast iron.

Graphitic corrosion in regions subject to high fluid velocities or cavitation may be misdiagnosed as simple erosion or cavitation. Graphitic corrosion produces a relatively soft corrosion product that is susceptible to erosion at significantly lower fluid velocities and to cavitation at significantly lower intensities. The continual removal of the corrosion product may obscure the evidence of active graphitic corrosion. In such cases, a metallographic examination may be required to determine if graphitic corrosion is a contributing factor (Case History 12.1).

Related Problems

See also Chap. 16, "Galvanic Corrosion."

CASE HISTORY 17.1

Industry:	Metals
Specimen Location:	Subterranean cooling water supply line
Specimen Orientation:	Horizontal
Environment:	Internal: Cooling water
	External: Soil
Time in Service:	25 years
Sample Specifications:	10 in. (25 cm) outside diameter, gray cast iron pipe

Figure 17.7 shows a 10 ft (3 m) section of a main cooling water line that fractured along its bottom. Close examination of this photograph will disclose that this fracture runs the entire length of the section. The pipe wall had been converted from metallic cast iron to a soft, brittle, black corrosion product along a narrow zone that ran the length of the pipe. Note in Figs. 17.8 and 17.5 that the corrosion product has the same contour as the original pipe. Striking the affected zone with a hammer caused the corrosion product to shatter, revealing a thin layer of intact metal below.

Internal surfaces were coated with a thin layer of reddish iron oxides. Significant corrosion was not observed on this surface.

Microstructural examinations of the external surface revealed an interconnecting network of graphite flakes embedded in a matrix of iron oxide.

Heavy equipment had passed over the site of failure initiation shortly before the rupture was discovered. It is probable that stresses associated with the heavy equipment initiated a fracture in the severely corroded pipe bottom. Once the fracture initiated, it propagated down the length of the line in response to stresses imposed by internal pressure.

Figure 17.7 Section of longitudinally fractured subterranean cooling water line. Note that the crack runs the entire length of the line.

Figure 17.8 The black outer covering is corrosion product; the reddish-brown surface is coated with air-formed iron oxide.

CASE HISTORY 17.2

Industry:	Metals
Specimen Location:	Pump impeller
Specimen Orientation:	Vertical
Environment:	Cooling water: 80–100°F (27–38°C), calcium 780 ppm, sodium 460 ppm, magnesium 220 ppm, chloride 1200 ppm, sulfate 300 ppm, pH 4.1, conductivity 3500 μmhos/cm
Time in Service:	3 months
Sample Specifications:	13 in. (33 cm) outside diameter, gray cast iron pump impeller

The impeller shown in Fig. 17.9 was removed from a cooling water return line associated with a large press. Failure of these impellers was a chronic problem, requiring impeller replacement every 3 or 4 months.

Examination revealed severe metal loss on all surfaces, but it was worst on the outer edges of the vanes. Deep grooves had formed near the hub between the spiral vanes.

Most of the surface is covered with a black corrosion product that is thicker in relatively low-flow areas near the hub. This layer of soft corrosion product can be shaved from corroded surfaces. Microstructural examinations revealed flakes of graphite embedded in iron oxide near the surfaces.

Turbulence and high fluid velocities resulting from normal pump operation accelerated metal loss by abrading the soft, graphitically corroded surface (erosion-corrosion). The relatively rapid failure of this impeller is due to the erosive effects of the high-velocity, turbulent water coupled with the aggressiveness of the water. Erosion was aided in this case by solids suspended in the water.

Figure 17.9 Pump impeller that has suffered erosion associated with graphitic corrosion.

CASE HISTORY 17.3

Industry:	Metals
Specimen Location:	Pump housing, quench-system water supply
Specimen Orientation:	Horizontal
Environment:	Untreated cooling water, pH 6
Time in Service:	Unknown
Sample Specifications:	Gray cast iron

Figure 17.10 shows metal loss on the throat of the pump housing. External pump housing surfaces were also affected (Fig. 17.11). Note the large tubercles. (Tubercles are knoblike mounds of corrosion products. They typically have a hard, black outer shell enclosing porous reddish-brown or black iron oxides) (see Chap. 3, "Tuberculation"). The metal surface beneath these tubercles had sustained graphitic corrosion, in some cases to a depth of ¼ in. (0.6 cm) (Fig. 17.12).

Internal surfaces of the pump show severe wastage (Fig. 17.13). The wasted region is free of corrosion products except for a small amount of soft, black material. Metal loss in this area was as deep as ½ in. (1.3 cm). The reddish coating partially covering the smooth area above the wasted zone in Fig. 17.13 was applied to mitigate corrosion. Where this coating is intact, corrosion has not occurred. Where it is absent, attack is severe (Fig. 17.14).

The pump has experienced graphitic corrosion. Figures 17.10, 17.12, and 17.14 illustrate typical appearances of graphitically corroded cast iron. In addition, cavitation damage (see Chap. 12) has produced severe metal loss in specific areas (see Fig. 17.13). The soft, friable corrosion products produced by graphitic corrosion are susceptible to cavitation damage at relatively low levels of cavitation intensity.

Figure 17.10 Appearance of the graphitically corroded throat of a pump housing.

Figure 17.11   External surface of a pump housing showing tubercles capping sites of graphitic corrosion.

Figure 17.12   Corroded area beneath a tubercle on the external surface. The shiny black corrosion product has been partially removed to reveal the depth of penetration.

Figure 17.13 Severe metal loss along an internal surface.

Figure 17.14 Cross section through the pump housing wall. Note the black and brown graphitic corrosion-product layer near the center of the photo.

CASE HISTORY 17.4

Industry:	Refining
Specimen Location:	Impeller from a recirculating pump in a cooling water system
Specimen Orientation:	Vertical
Environment:	Treated cooling water
Time in Service:	Unknown
Sample Specifications:	Gray cast iron

The specimen in Fig. 17.15 is part of a cast iron pump impeller. The spongelike surface contours are apparent, as is the black coating that covers surfaces exposed to the cooling water. Microstructural examinations revealed preferential deterioration of the iron matrix surrounding the graphite flakes.

Graphitic corrosion of the cast iron produced a soft, mechanically weak corrosion product that is highly susceptible to cavitation damage, even at relatively low cavitation intensities. The black coating on the impeller surface is visual evidence of graphitic corrosion. The spongelike surface contours are typical of cavitation damage (see Chap. 12).

Figure 17.15 Section of a cast iron pump impeller that has suffered graphitic corrosion followed by cavitation damage.

CASE HISTORY 17.5

Industry:	Chemical process
Specimen Location:	Well pump
Specimen Orientation:	Vertical
Environment:	Raw well water, temperature 55°F (13°C), pH 7, pressure 200 psi (1.4 MPa)
Time in Service:	5½ years
Sample Specifications:	Gray cast iron

The well supplies cooling water to the plant. During operation of the system, a reduction in flow rate was noticed. The well was shut down and the pump inspected.

Inspection revealed severe metal loss on internal surfaces at the base of the vanes resulting in a large, irregular perforation (Figs. 17.16 through 17.18). In addition, metal loss was observed in a distinct circumferential zone adjacent to internal threads (Fig. 17.17).

Metal loss in these areas had produced a smooth surface, free of deposits and corrosion products. The rest of the internal surface was covered by a thin, uniform layer of soft, black corrosion product. The graphitically corroded surfaces of the pump casing provided soft, friable corrosion products that were relatively easily dislodged by the abrasive effects of high-velocity or turbulent water (erosion-corrosion).

Failures of this type occurred every 6 to 9 years. The well was operated intermittently, since several available wells are operated on a rotating basis. It is probable that most of the graphitic corrosion occurred during idle times; actual metal loss occurred during operation of the pump.

Figure 17.16 Pump casing as received. Note the large perforation near the center.

Figure 17.17 Longitudinally cut half-section. Note the severe wall thinning and the perforation of the wall on the left side. Compare it to the wall thickness on the right side. Also note the circumferential band of irregular metal loss just above the threads, near the bottom of the casing.

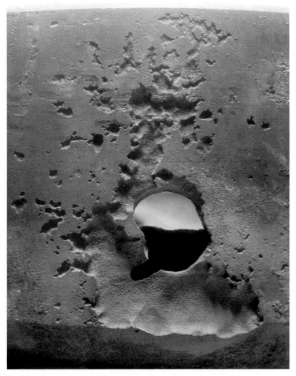

Figure 17.18 Close-up of the perforation in Fig. 17.16 viewed from the external surface. Graphitic corrosion of the external surface permitted erosion by turbulent water escaping through the perforation.

Glossary*

acid   A compound that yields hydrogen ions (H^+) when dissolved in water.

active metal   A metal ready to corrode or being corroded. Contrast with noble metal.

admiralty brass   A series of three zinc alloys of copper containing 1% tin and one of three inhibitors for dezincification; hence, arsenical admiralty, antimonial admiralty, and phosphorized admiralty. Typical applications for this material are condenser, evaporator and heat exchanger tubing, condenser tubing plates, and ferrules.

alkaline   1. Having properties of an alkali.   2. Having a pH greater than 7.

alloy   A substance having metallic properties and being composed of two or more chemical elements of which at least one is a metal.

alloy steel   Steel containing specified quantities of alloying elements added to effect changes in mechanical or physical properties.

alloy system   A family of alloys having in common a single, compositionally predominant metal.

aluminum brass   A zinc alloy of copper containing 2% aluminum and 0.1% arsenic for inhibition of dezincification. Typical applications for this material are condenser, evaporator and heat exchanger tubing, condenser tubing plates, and ferrules. It is often specified when erosion resistance exceeding that of normal brasses is required.

amphoteric   Capable of reacting chemically either as acid or a base. In reference to certain metals, signifies their propensity to corrode at both high and low pH.

annealing   A generic term denoting a treatment consisting of heating to and holding at a suitable temperature followed by cooling at a suitable rate used primarily to soften metallic materials, but also to produce desired changes in other properties or in microstructure simultaneously. When the term is used by itself, full annealing is implied. When the treatment is applied only for the relief of stress, it is properly called stress-relieving or stress-relief annealing.

---

* Portions of the glossary are from the *Metals Handbook*, vol. 13 and the *Metals Handbook*, desk edition. Reprinted with permission from American Society for Metals.

anode   The electrode of an electrolyte cell at which oxidation occurs. Electrons flow away from the anode in the external circuit. It is usually at the electrode that corrosion occurs and metal ions enter solution. Contrast with cathode.

anodizing   Forming a conversion coating on a metal surface by anodic oxidation; most frequently applied to aluminum.

applied stress   Stresses extrinsic to the material itself, such as those imposed by service conditions. Contrast with residual stress.

austenitic stainless steel   A nonmagnetic stainless steel possessing a microstructure of austenite. In addition to chromium, these steels commonly contain at least 8% nickel.

autocatalytic corrosion   Corrosion processes that are self-sustaining.

base metal   1. In welding, the metal to be welded.   2. After welding, that part of the metal that was not melted.

brass   An alloy consisting mainly of copper (over 50%) and zinc to which smaller amounts of other elements may be added.

brittle fracture   Separation of a solid accompanied by little or no macroscopic plastic deformation.

bronze   A copper-rich copper–tin alloy with or without small proportions of other elements such as zinc and phosphorus. Also, certain other essentially binary copper-based alloys containing no tin, such as aluminum bronze, silicon bronze, and beryllium bronze.

cathode   In a corrosion cell, the area over which reduction is the principal reaction. It is usually an area that is not attacked.

cathodic protection   Partial or complete protection of a metal from corrosion by making it a cathode, using either a galvanic or an impressed current.

caustic   A hydroxide of a light metal, such as sodium hydroxide or potassium hydroxide.

cavitation   The formation and instantaneous collapse of innumerable tiny voids or cavities within a liquid subjected to rapid and intense pressure changes.

cavitation damage   The degradation of a solid body resulting from its exposure to cavitation. This may include loss of material, surface deformation, or changes in properties or appearance.

cementite   A compound of iron and carbon, known chemically as iron carbide and having the approximate chemical formula Fe_3C.

chevron pattern   A fractographic pattern of radial marks that resemble nested letters V. Chevron patterns are typically found on brittle fracture surfaces in parts whose widths are considerably greater than their thicknesses. The points of the chevrons can be traced back to the fracture origin.

closed recirculating   A cooling system in which water is recirculated in a closed loop. Such systems experience little, if any, exposure to the atmosphere.

cold flow   Deformation of an elastic material resulting from stresses applied at ambient temperatures.

cold work   Permanent deformation of a metal produced by an external force.

concentration cell   An electrolytic cell, the electromotive force of which is caused by a difference in concentration of some component in the electrolyte. This difference leads to the formation of discrete cathode and anode regions.

corrosion   The chemical or electrochemical reaction between a material, usually a metal, and its environment that produces a deterioration of the material and its properties.

corrosion fatigue   The process in which a metal fractures prematurely under conditions of simultaneous corrosion and repeated cyclic loading at lower stress levels or fewer cycles than would be required in the absence of the corrosive environment.

corrosion product   Substance formed as a result of corrosion.

corrosivity   The tendency of an environment to cause corrosion in a given corrosion system.

crevice corrosion   A type of concentration cell corrosion; corrosion caused by the concentration or depletion of dissolved salts, metal ions, oxygen or other gases, and such, in crevices or pockets remote from the principal fluid stream, with a resultant buildup of differential cells that ultimately cause deep pitting.

cupronickel   A copper-based alloy containing 5 to 30% nickel.

cyclic stress   A stress whose magnitude fluctuates. Contrast with static stress.

dealloying (see also **selective leaching**)   The selective corrosion of one or more components of a solid solution alloy. Also called *parting* or *selective leaching.*

defect   An imperfection in a material that contributes significantly to failure or limited serviceability. Contrast with flaw.

denickelification   Corrosion in which nickel is selectively leached from nickel-containing alloys. Most commonly observed in copper–nickel alloys after extended service in fresh water.

depassivation (activation)   The changing of a metal surface from a chemically nonreactive (passive) condition to a reactive condition.

depassivation agent   A substance, usually an ion such as chloride, whose reaction with a metal surface destroys its passive character.

dezincification   Corrosion in which zinc is selectively leached from zinc-containing alloys. Most commonly found in copper–zinc alloys containing less than 85% copper after extended service in water containing dissolved oxygen. Uniform loss of zinc is termed *layer-type dezincification;* localized loss of zinc is termed *plug-type dezincification.*

ductile fracture   Fracture characterized by tearing of metal accompanied by appreciable gross plastic deformation and expenditure of considerable energy.

ductility   The ability of a material to deform plastically without fracturing.

eddy-current testing   An electromagnetic nondestructive testing method in which eddy-current flow is induced in the test object. Changes in flow caused by variations in the object are reflected into a nearby coil or coils where they are detected and measured by suitable instrumentation.

electrolyte   1. An ionic conductor.   2. A liquid, most often a solution, that will conduct an electric current.

erosion   Destruction of metals or other materials by the abrasive action of moving fluids, usually accelerated by the presence of solid particles or matter in suspension. When corrosion occurs simultaneously, the term *erosion-corrosion* is often used.

exfoliation   A type of corrosion that progresses approximately parallel to the outer surface of the metal, causing layers of the metal or its oxide to be elevated by the formation of corrosion products.

failure   A general term used to imply that a part in service (1) has become completely inoperable, (2) is still operable but is incapable of satisfactorily performing its intended function, or (3) has deteriorated seriously to the point that it has become unreliable or unsafe for continued use.

fatigue   The phenomenon leading to fracture under repeated or fluctuating mechanical stresses having a maximum value less than the tensile strength of the material.

fatigue limit   The maximum stress that presumably leads to fatigue fracture in a specified number of stress cycles.

ferrite   Designation commonly assigned to alpha iron-containing elements in solid solution.

ferrous hydroxide   A white corrosion product of iron, $Fe(OH)_2$. Ferric ion incorporated in the substance will alter the color to green, brown, or black.

flaw   As used in this book, an imperfection in a material that does not affect its usefulness or serviceability. Contrast with defect.

flow-induced vibration   Tube vibration resulting from the mechanical effects of fluids impinging on tube surfaces. Such vibrations may induce wear or cracking, especially if the tubes vibrate at their natural frequency.

flux   In welding, material used to prevent the formation of or to dissolve and facilitate the removal of oxides and other undesirable substances.

fretting   A type of wear that occurs between tight fitting surfaces subjected to cyclic, relative motion of extremely small amplitude. Usually, fretting is accompanied by corrosion, especially of the very fine wear debris.

galvanic cell   A cell in which chemical change is the source of electrical energy. It usually consists of two dissimilar conductors in contact with each other and with an electrolyte or of two similar conductors in contact with each other and with dissimilar electrolytes.

galvanic corrosion   Corrosion associated with the current of a galvanic cell consisting of two dissimilar conductors in an electrolyte or two similar conductors in dissimilar electrolytes.

galvanic potential   The magnitude of the driving force in an electrochemical reaction resulting from the coupling of dissimilar materials exposed to a common, corrosive environment.

galvanic series   A series of metals and alloys arranged according to their relative electrode potentials in a specified environment.

gas porosity   Fine holes or pores within a metal that are caused by entrapped gas or by evolution of dissolved gas during solidification.

grain   An individual crystal in a polycrystalline metal or alloy.

grain boundary   A narrow zone in a metal corresponding to the transition from one crystallographic orientation to another, thus separating one grain from another.

graphitic corrosion   Corrosion of gray or nodular cast iron in which the iron matrix is selectively leached away, leaving a porous mass of graphite behind; it occurs in relatively mild aqueous solutions and on buried pipe fittings.

graphitization   A metallurgical term describing the formation of graphite in iron or steel, usually from decomposition of iron carbide at elevated temperatures. Not recommended as a term to describe graphitic corrosion.

heat-affected zone   That portion of the base metal that was not melted during welding but whose microstructure and mechanical properties were altered by the heat.

hematite   A magnetic form of iron oxide, Fe_2O_3. Hematite is gray to bright red. The reddish forms are nonprotective and indicate the presence of high levels of oxygen.

holidays   Discontinuities in a coating (such as porosity, cracks, gaps, and similar flaws) that allow areas of base metal to be exposed to any corrosive environment that contacts the coated surface.

hydrated ferric oxide (ferric hydroxide)   A flaky, red to brown corrosion product of iron or steel that forms upon exposure to subterranean, atmospheric, or aqueous environments, $Fe_2O_3 \cdot 3H_2O$.

hydrolysis   A chemical process of decomposition involving splitting of a bond and addition of the elements of water.

hydrous ferrous ferrite   The hydrated form of ferrosoferric oxide, Fe_3O_4, a black, magnetic corrosion product, $Fe_3O_4 \cdot nH_2O$. It is oxidized to hematite, Fe_2O_3, when heated in air.

hydrous ferrous oxide   The hydrated form of ferrous oxide, FeO, a jet black corrosion product, $FeO \cdot nH_2O$. It is readily oxidized by air.

intergranular   Between crystals or grains. Also called intercrystalline.

intergranular corrosion   Corrosion occurring preferentially at grain boundaries, usually with slight or negligible attack on the adjacent grains.

laminar flow   Streamlined flow in a fluid.

lap   A surface imperfection, with the appearance of a seam, caused by hot metal, fins, or sharp corners being folded over then being rolled or forged into the surface but without being welded.

local action   Corrosion caused by "local cells" on a metal surface.

local cells   A cell, the emf of which is due to differences of potential between areas on a metal surface in an electrolyte.

magnetite   A magnetic form of iron oxide, Fe_3O_4. Magnetite is dark gray to black and forms a protective film on iron surfaces.

matrix   The principal phase in which another constituent is embedded.

microbiologically influenced corrosion (MIC)   Deterioration of metals as a result of the metabolic activities of microorganisms.

microstructure   The structure of a metal as revealed by microscopic examination of the etched surface of a polished specimen.

mild steel   Carbon steel having a maximum carbon content of approximately 0.25%.

Neumann band   Mechanical twin in ferrite.

noble metal   A metal with marked resistance to chemical reaction, particularly to oxidation and to solution by inorganic acids. Contrast with active metal.

once-through   A cooling system in which water is received from the plant supply, passed through the process equipment, and returned to a receiving body of water.

open recirculating   A cooling system in which water is taken from a cooling tower or evaporation pond, passes through the process equipment, then returns to the evaporation unit. This system requires fresh makeup water to replenish that lost by evaporation and blowdown.

parting   See **dealloying**.

passivation   The changing of a chemically active surface of a metal to a much less reactive state.

passivity   A condition in which a metal, because of an impervious covering of oxide or other compound, has a potential much more positive than that of the metal in its active state.

pearlite   A microstructural aggregate consisting of alternate lamellae of ferrite and cementite.

penetration   In welding, the distance from the original surface of the base metal to that point at which fusion ceased.

pH   The negative logarithm of the hydrogen ion activity; it denotes the degree of acidity or basicity of a solution. At 77°F (25°C), 7.0 is the neutral value.

Decreasing values below 7.0 indicate increasing acidity; increasing values above 7.0 indicate increasing basicity.

pipe   The central cavity formed by contraction in metal, especially ingots, during solidification.

pit   A distinct cavity in a metal surface resulting from highly localized corrosion.

pitting   Forming small sharp cavities in a metal surface by corrosion.

plain carbon steel (ordinary steel)   Steel containing carbon up to about 2% and only residual quantities of other elements except those added for deoxidation.

plankton   Minute animal and plant life found in water.

plate-and-frame heat exchanger   An exchanger that consists of a stack of thin plates supported in a frame. The plates are typically corrugated. The two fluids flow along opposite sides of each plate.

radiographic inspection   A nondestructive testing technique that uses penetrating radiation (x-rays, gamma rays, etc.) to record differences in material density, thickness, and composition on film or other forms of radiation detectors.

residual stress   Stresses that remain within a body as a result of plastic deformation.

root crack   A crack in either the weld or heat-affected zone at the root of a weld.

root of joint   The portion of a weld joint where the members are closest to each other before welding. In cross section, this may be a point, a line, or an area.

root of weld   The points at which the weld bead intersects the base-metal surfaces either nearest to or coincident with the root of joint.

rust   A corrosion product consisting of hydrated oxides of iron.

seam   On a metal surface, an unwelded fold or lap that appears as a crack, usually resulting from a discontinuity.

seam welding   Making a longitudinal weld in sheet metal or tubing.

selective leaching   Corrosion in which one element is preferentially removed from an alloy, leaving a residue (often porous) of the elements that are more resistant to the particular environment.

sensitization   In austenitic stainless steels, the precipitation of chromium carbides, usually at grain boundaries, on exposure to temperatures of about 1000 to 1500°F (550 to 850°C), leaving the grain boundaries depleted of chromium and therefore susceptible to preferential attack by a corroding medium.

sessile   Permanently attached; not free to move.

shell-and-tube heat exchangers   The most widely used form of heat exchanger in which one fluid passes through a number of tubes housed in a shell. The second fluid passes through the shell.

siderite   Ferrous carbonate, $FeCO_3$.

siphonic gas exsolution   The formation of gas bubbles in a fluid caused by decreasing fluid pressure associated with flow.

spalling   The cracking and flaking of particles out of a surface.

stagnant   The condition of being motionless.

stainless steel   Any of several steels containing 12 to 30% chromium as the principal alloying element; the steels usually exhibit passivity in aqueous environments.

static stress   A stress whose magnitude remains at a constant value. Contrast with cyclic stress.

stress   Force per unit area, often thought of as force acting through a small area within a plane. It can be divided into components, normal and parallel to the plane, called *normal stress* and *shear stress,* respectively. *True stress* denotes the stress where force and area are measured at the same time. *Conventional stress,* as applied to tension and compression tests, is force divided by original area.

stress-corrosion cracking   Failure by cracking under the combined action of a specific corrosive and stress, either external (applied) stress or internal (residual) stress. Cracking may be either intergranular or transgranular, depending on the metal and the corrosive medium.

stress frequency   The number of times a stress cycle is repeated in a unit of time.

stress raisers   Changes in contour or discontinuities in structure that cause local increases in stress.

synergism   Cooperative action of discrete agencies such that the total effect is greater than the sum of the effects taken independently.

tensile strength   In tensile testing, the ratio of maximum load to original cross-sectional area. Also called *ultimate strength.*

transgranular   Through or across crystals or grains. Also called *intracrystalline* or *transcrystalline.*

tubercle   A knoblike structure of corrosion products that forms over corrosion sites on iron-based metals.

tuberculation   The formation of localized corrosion products in the form of knoblike mounds called *tubercles.*

twin   Two portions of a crystal having a definite crystallographic relationship. Twins can be thermally or mechanically produced.

ultrasonic testing   A nondestructive test applied to sound-conductive materials having elastic properties for the purpose of locating inhomogeneities or structural discontinuities within a material by means of an ultrasonic beam.

underbead crack   A subsurface crack in the base metal near a weld.

weld   A union made by welding.

weld bead   A deposit of filler metal from a single welding pass.

welding current   The current flowing through a welding circuit during the making of a weld.

weldment   An assembly whose component parts are joined by welding.

weld metal   That portion of a weld that has been melted during welding.

yield strength   The stress at which a material exhibits a specified deviation from proportionality of stress and strain. Compare with tensile strength.

Further Reading

Anderson, R. C., *Inspection of Metals,* vol. I, Visual Examination, American Society for Metals, Metals Park, Ohio, 1983.

Butler, G. and H. C. K. Ison, *Corrosion and its Prevention in Waters,* Robert E. Krieger Publishing Company, Huntington, New York, 1978.

"Corrosion," *Metals Handbook,* edition 9, vol. 13, American Society for Metals, Metals Park, Ohio, 1987.

Corrosion Basics, National Association of Corrosion Engineers, Houston, Texas, 1984.

Dexter, S. C. (ed.), *Biologically Induced Corrosion,* Proceedings of the International Conference on Biologically Induced Corrosion, National Association of Corrosion Engineers, Houston, Texas, 1986.

Dillon, C. P. (ed.), *Forms of Corrosion, Recognition and Prevention,* NACE Handbook I, National Association of Corrosion Engineers, Houston, Texas, 1982.

During, E. D. D., *Corrosion Atlas,* vol. 1, Elsevier Science Publishing Company, Inc., New York, 1988.

Evans, U. R., *The Corrosion and Oxidation of Metals, First Supplementary Volume,* Edward Arnold Ltd., London, 1968.

Evans, U. R., *Corrosion and Oxidation of Metals: Scientific Principles and Practical Applications,* Edward Arnold Ltd., London, 1960.

"Failure Analysis and Prevention," *Metals Handbook,* edition 8, vol. 10, American Society for Metals, Metals Park, Ohio.

Fontana, M. G., and N. D. Green, *Corrosion Engineering,* McGraw-Hill, Inc., New York, 1967.

Gilbert, R. J., and D. W. Lovelock (eds.), *Microbial Aspects of the Deterioration of Materials,* Academic Press, London, 1975.

Internal Corrosion of Water Distribution Systems, Cooperative Research Report, AWWA Research Foundation, Denver, Colorado, DVGW-Forschungsstelle, 1985.

Kemmer, F. N. (ed.), *The Nalco Water Handbook,* McGraw-Hill Book Company, New York, 1979.

Logan, H. L., *The Stress Corrosion of Metals,* National Bureau of Standards, Washington, D.C., John Wiley and Sons, Inc., New York, 1966.

McCall, J. L., and P. M. French (eds.), *Metallography in Failure Analysis,* Plenum Press, New York, 1978.

McCoy, J. W., *The Chemical Treatment of Cooling Water,* 2nd ed., Chemical Publishing Co., New York, 1983.

Shreir, L. L. (ed.), *Corrosion,* George Newness, Ltd., Tower House, London, 1963.

Speller, F. N., *Corrosion / Causes and Prevention,* McGraw-Hill, Inc., New York, 1951.

Thielsch, H., *Defects and Failures in Pressure Vessels and Piping,* Reinhold Publishing Corp., New York, 1965.

Uhlig, H. H., *Corrosion and Corrosion Control,* John Wiley and Sons, Inc., New York, 1963.

Uhlig, H. H. (ed.), *The Corrosion Handbook,* John Wiley and Sons, Inc., New York, 1948.

Industry Index

Air conditioning industry:
  cavitation damage in, 289–291
  material defects in, 322–323
Air conditioning system, 2–3
Alcohol production industry, biologically
  influenced corrosion in, 150–152
Aluminum industry, acid corrosion in,
  180–181
Automotive industry:
  cavitation damage in, 281–284
  weld defects in, 353–354

Building industry:
  erosion-corrosion in, 263
  underdeposit corrosion in, 94–95

Chemical process industry:
  cavitation damage in, 292–293
  crevice corrosion in, 33–34
  erosion-corrosion in, 268–269
  galvanic corrosion in, 368–369
  graphitic corrosion in, 387–390
  heat exchangers in, 1
  material defects in, 319–322
  oxygen corrosion in, 112–113, 115–116
  stress-corrosion cracking in, 210–221

Hospital industry, erosion-corrosion in,
  264–266
Hotel industry, dealloying in, 310–311

Metals industry:
  acid corrosion in, 177–178
  alkaline corrosion in, 196–197
  biologically influenced corrosion in,
    156–157
  corrosion fatigue in, 237–238
  crevice corrosion in, 34
  graphitic corrosion in, 381–387
  oxygen corrosion in, 117–118
  tuberculation in, 63
  underdeposit corrosion in, 91–92
  weld defects in, 350–353
Mining industry, dealloying in, 304–305
Municipal cooling system, galvanic corrosion in, 367

Natural gas industry, underdeposit corrosion in, 95–96
Nuclear utility industry:
  biologically influenced corrosion in, 135,
    143–144, 147–148, 154–156

Nuclear utility industry (*Cont.*):
  crevice corrosion in, 35
  tuberculation in, 64, 65

Polyethylene film industry, alkaline corrosion, 195–196
Power generation industry *see* Utility
  industry
Primary metals industry *see* Metals
  industry
Pulp and paper industry:
  acid corrosion in, 182–183
  cavitation damage in, 285–286
  erosion-corrosion in, 261–262
  galvanic corrosion in, 370–371

Refinery industry:
  corrosion fatigue in, 235–237
  dealloying in, 305–306
  graphitic corrosion in, 388
  heat exchangers in, 1
  oxygen corrosion in, 113–114
  underdeposit corrosion in, 86–87,
    90–91, 93

Steel industry:
  acid corrosion in, 178–180
  biologically influenced corrosion in,
    149–150
  cavitation damage in, 287–289
  crevice corrosion in, 31–33
  erosion-corrosion in, 266–268
  material defects in, 325–326
  stress-corrosion cracking in, 223
  tuberculation in, 58

Utility industry:
  alkaline corrosion in, 198
  biologically influenced corrosion in,
    152–153
  corrosion fatigue in, 233–235
  dealloying in, 307–309
  erosion-corrosion in, 251–260
  material defects in, 323–325
  stress-corrosion cracking in, 222
  tuberculation in, 60–62
  underdeposit corrosion in, 88–90
  weld defects in, 346–349
  See also Natural gas industry; Nuclear
    utility industry

Metallurgy Index

Water Treatment Index

Subject Index